Mathematics Standard Level

for the IB Diploma

Exam Preparation Guide

Paul Fannon, Vesna Kadelburg,
Ben Woolley, Stephen Ward

CAMBRIDGE
UNIVERSITY PRESS

CAMBRIDGE
UNIVERSITY PRESS

University Printing House, Cambridge CB2 8BS, United Kingdom

One Liberty Plaza, 20th Floor, New York, NY 10006, USA

477 Williamstown Road, Port Melbourne, VIC 3207, Australia

314–321, 3rd Floor, Plot 3, Splendor Forum, Jasola District Centre,
New Delhi – 110025, India

79 Anson Road, #06–04/06, Singapore 079906

Cambridge University Press is part of the University of Cambridge.

It furthers the University's mission by disseminating knowledge in the pursuit of
education, learning and research at the highest international levels of excellence.

Information on this title: education.cambridge.org

© Cambridge University Press 2014

First published 2014
20 19 18 17 16 15 14 13 12 11 10 9 8

Printed in Great Britain by CPI Group (UK) Ltd, Croydon CRO 4YY

A catalogue record for this publication is available from the British Library

ISBN 978-1-107-65315-3 Paperback

Cover image: David Robertson/Alamy

CONTENTS

INTRODUCTION

ABOUT THIS BOOK

If you are using this book, you're probably getting quite close to your exams. You may have started off as a bright-eyed student keen to explore international perspectives in mathematics and the nature of mathematical knowledge, but now you want to know how to get the best possible grade! This book is designed to revise the entire core material that you need to know, and to provide examples of the most common types of exam questions for you to practise, along with some hints and tips regarding exam technique and common pitfalls.

> ⚠ Any common pitfalls and useful exam hints will be highlighted in these boxes.

> 🖩 This type of box will be used to point out where graphical calculators can be used effectively to simplify a question or speed up your work. Common calculator pitfalls will also be highlighted in such boxes.

> ▶▶ If the material in a chapter involves maths outside of that chapter, this kind of box will direct you to the relevant part of the book where you can go and remind yourself of that maths.

The most important ideas and formulae are emphasised in the 'What you need to know' sections at the start of each chapter. When a formula or set of formulae is given in the Formula booklet, there will be a book icon next to it . If formulae are not accompanied by such an icon, they do not appear in the Formula booklet and you may need to learn them or at least know how to derive them.

For Mathematics Standard Level, each of the written papers:

- is worth 40% of the final grade
- is one and a half hours long (plus 5 minutes of reading time)
- has a total of 120 marks available
- contains 9 or 10 short questions and 3 or 4 long questions.

The difference between the two papers is that calculators are **not allowed** for Paper 1, but are required for Paper 2.

 In this book questions which are designed to be done without a calculator are accompanied by a non-calculator icon.

 Questions which are only sensible to do with a calculator are marked with a calculator icon.

All other questions should be attempted first without a calculator, and then with one.

Between Papers 1 and 2 the majority of the material in the Standard Level course will be assessed.

IMPORTANT EXAM TIPS

- **Grab as many marks as you can.**
  - If you cannot do an early part of a question, write down a sensible answer and use that in later parts or, if the part you could not do was a 'show that' task, use the given result. You will still pick up marks.
  - Do not throw away 'easy marks':
    - Give all answers exactly or to three significant figures. Each time you fail to do so you will lose a mark.
    - Do not use rounded intermediate values, as this can result in an inaccurate answer; store all intermediate values in your calculator.
    - Read the questions carefully and make sure you have provided the answer requested. For example, if the question asks for coordinates, give both x and y values. If the question asks for an equation, make sure that you have something with an equals sign in it.
- **The questions are actually worded to help you.**
  - Make sure you know what each command term means. (These are explained in the IB syllabus.) For example:
    - 'Write down' means that there does not need to be any working shown. So, for this type of question, if you are writing out lines and lines of algebra, you have missed something.
    - 'Hence' means that you have to use the previous part somehow. You will not get full marks for a correct answer unless you explicitly show how you have used the previous part.
    - 'Hence or otherwise' means that you can use any method you wish, but it's a pretty big hint that the previous part will help in some way.
    - 'Sketch' means that you do not need to do a precise and to-scale drawing; however, you should label all the important points – at the very least where the curve crosses any axes.
  - If the question refers to solutions, you should expect to get more than one answer.
  - Look out for links between the parts of a question, particularly in the long questions.
- **Use your 5 minutes of reading time effectively.**
  - Decide on the order in which you will attempt the questions. It is very rarely most efficient to answer the questions in the order in which they are set; in particular, the last few short questions often involve a lot of work relative to the number of marks available.

 Make sure you leave enough time for the long questions, some parts of which can be fairly straightforward.

  - For Paper 2, decide which questions can be done easily or checked effectively on the calculator. Do not be surprised if this is the majority of questions.
  - Try to classify which section of the course each question is about.

 Practise using the reading time when attempting your practice papers.

Most importantly, there is nothing like good preparation to make you feel relaxed and confident going into an exam, which will help you achieve your best possible result.

Good luck!
The author team

1 EXPONENTS AND LOGARITHMS

WHAT YOU NEED TO KNOW

- The rules of exponents:

  - $a^m \times a^n = a^{m+n}$

  - $\dfrac{a^m}{a^n} = a^{m-n}$

  - $(a^m)^n = a^{mn}$

  - $a^{\frac{m}{n}} = \sqrt[n]{a^m}$

  - $a^{-n} = \dfrac{1}{a^n}$

  - $a^n \times b^n = (ab)^n$

  - $\dfrac{a^n}{b^n} = \left(\dfrac{a}{b}\right)^n$

- The relationship between exponents and logarithms:

  - $a^x = b \iff x = \log_a b$ where a is called the base of the logarithm

  - $\log_a a^x = x$

  - $a^{\log_a x} = x$

- The rules of logarithms:

  - $\log_c a + \log_c b = \log_c ab$

  - $\log_c a - \log_c b = \log_c \dfrac{a}{b}$

  - $\log_c a^r = r \log_c a$

  - $\log_c \left(\dfrac{1}{a}\right) = -\log_c a$

  - $\log_c 1 = 0$

- The change of base rule: $\log_b a = \dfrac{\log_c a}{\log_c b}$

- There are two common abbreviations for logarithms to particular bases:

  - $\log_{10} x$ is often written as $\log x$

  - $\log_e x$ is often written as $\ln x$

- The graphs of exponential and logarithmic functions:

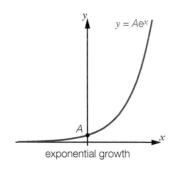

exponential growth

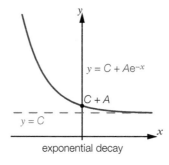

exponential decay

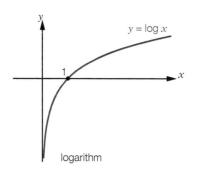

logarithm

⚠ EXAM TIPS AND COMMON ERRORS

- You must know what you *cannot* do with logarithms:

  - $\log(x + y)$ cannot be simplified; it is **not** $\log x + \log y$

  - $\log(e^x + e^y)$ cannot be simplified; it is **not** $x + y$

  - $(\log x)^2$ is **not** $2\log x$, whereas $\log x^2 = 2\log x$

  - $e^{2+\log x} = e^2 e^{\log x} = e^2 x$ **not** $e^2 + x$

1.1 SOLVING EXPONENTIAL EQUATIONS

WORKED EXAMPLE 1.1

Solve the equation $4 \times 5^{x+1} = 3^x$, giving your answer in the form $\dfrac{\log a}{\log b}$.

$\log(4 \times 5^{x+1}) = \log(3^x)$

$\Leftrightarrow \log 4 + \log(5^{x+1}) = \log(3^x)$

> Since the unknown is in the power, we take logarithms of each side.
>
> We then use the rules of logarithms to simplify the expression. First use $\log(ab) = \log a + \log b$

 A common mistake is to say that $\log(4 \times 5^{x+1}) = \log 4 \times \log(5^{x+1})$.

$\Leftrightarrow \log 4 + (x+1)\log 5 = x \log 3$

> We can now use $\log a^k = k \log a$ to get rid of the powers.

$\Leftrightarrow \log 4 + x \log 5 + \log 5 = x \log 3$

$\Leftrightarrow x \log 3 - x \log 5 = \log 4 + \log 5$

$\Leftrightarrow x(\log 3 - \log 5) = \log 4 + \log 5$

> Expand the brackets and collect the terms containing x on one side.

$\Leftrightarrow x = \dfrac{\log 4 + \log 5}{\log 3 - \log 5}$

$\Leftrightarrow x = \dfrac{\log 20}{\log\left(\dfrac{3}{5}\right)}$

> Use the rules of logarithms to write the solution in the correct form:
>
> $\log a + \log b = \log ab$
>
> $\log a - \log b = \log\left(\dfrac{a}{b}\right)$

Practice questions 1.1

1. Solve the equation $5^{3x+1} = 15$, giving your answer in the form $\dfrac{\log a}{\log b}$ where a and b are integers.

2. Solve the equation $3^{2x+1} = 4^{x-2}$, giving your answer in the form $\dfrac{\log p}{\log q}$ where p and q are rational numbers.

3. Solve the equation $3 \times 2^{x-3} = \dfrac{1}{5^{2x}}$, giving your answer in the form $\dfrac{\log p}{\log q}$ where p and q are rational numbers.

1.2 SOLVING DISGUISED QUADRATIC EQUATIONS

WORKED EXAMPLE 1.2

Find the exact solution of the equation $3^{2x+1} - 11 \times 3^x = 4$.

$3^{2x+1} - 11 \times 3^x = 4$

$\Leftrightarrow 3 \times 3^{2x} - 11 \times 3^x = 4$

$\Leftrightarrow 3 \times (3^x)^2 - 11 \times 3^x = 4$

We need to find a substitution to turn this into a quadratic equation.

First, express 3^{2x+1} in terms of 3^x:

$3^{2x+1} = 3^{2x} \times 3^1 = 3 \times (3^x)^2$

 Look out for an a^{2x} term, which can be rewritten as $(a^x)^2$.

Let $y = 3^x$. Then

$3y^2 - 11y - 4 = 0$

$\Leftrightarrow (3y+1)(y-4)$

$\Leftrightarrow y = -\dfrac{1}{3}$ or $y = 4$

After substituting y for 3^x, this becomes a standard quadratic equation, which can be factorised and solved.

▶ Disguised quadratic equations may also be encountered when solving trigonometric equations, which is covered in Chapter 5.

$\therefore 3^x = -\dfrac{1}{3}$ or $3^x = 4$

Remember that $y = 3^x$.

$3^x = -\dfrac{1}{3}$ is impossible since $3^x > 0$ for all x.

 With disguised quadratic equations, often one of the solutions is impossible.

$3^x = 4$

$\Leftrightarrow \log 3^x = \log 4$

$\Leftrightarrow x \log 3 = \log 4$

$\Leftrightarrow x = \dfrac{\log 4}{\log 3}$

Since x is in the power, we take logarithms of both sides. We can then use $\log a^k = k \log a$ to get rid of the power.

Practice questions 1.2

 4. Solve the equation $2^{2x} - 5 \times 2^x + 4 = 0$.

 5. Find the exact solution of the equation $e^x - 6e^{-x} = 5$.

 6. Solve the simultaneous equations $e^{x+y} = 6$ and $e^x + e^y = 5$.

1.3 LAWS OF LOGARITHMS

WORKED EXAMPLE 1.3

If $x = \log a$, $y = \log b$ and $z = \log c$, write $2x + y - 0.5z + 2$ as a single logarithm.

$2\log a + \log b - 0.5\log c + 2$

$= \log a^2 + \log b - \log c^{0.5} + 2$

> We need to rewrite the expression as a single logarithm. In order to apply the rules for combining logarithms, each log must have no coefficient in front of it. So we first need to use $k\log x = \log x^k$.

$= \log a^2 b - \log c^{0.5} + 2$

$= \log\left(\dfrac{a^2 b}{\sqrt{c}}\right) + 2$

> We can now use $\log x + \log y = \log(xy)$ and $\log x - \log y = \log\left(\dfrac{x}{y}\right)$

$= \log\left(\dfrac{a^2 b}{\sqrt{c}}\right) + \log 100$

$= \log\left(\dfrac{100 a^2 b}{\sqrt{c}}\right)$

> We also need to write 2 as a logarithm so that it can then be combined with the first term. Since $10^2 = 100$, we can write 2 as $\log 100$.

 Remember that log on its own is taken to mean $\log_{10}$.

Practice questions 1.3

7. Given $x = \log a$, $y = \log b$ and $z = \log c$, write $3x - 2y + z$ as a single logarithm.

8. Given $a = \log x$, $b = \log y$ and $c = \log z$, find an expression in terms of a, b and c for $\log\left(\dfrac{10xy^2}{\sqrt{z}}\right)$.

9. Given that $\log a + 1 = \log b^2$, express a in terms of b.

10. Given that $\ln y = 2 + 4\ln x$, express y in terms of x.

 11. Consider the simultaneous equations
$$e^{2x} + e^y = 800$$
$$3\ln x + \ln y = 5$$
    (a) For each equation, express y in terms of x.
    (b) Hence solve the simultaneous equations.

1.4 SOLVING EQUATIONS INVOLVING LOGARITHMS

WORKED EXAMPLE 1.4

 Solve the equation $4\log_4 x = 9\log_x 4$.

$$\log_x 4 = \frac{\log_4 4}{\log_4 x} = \frac{1}{\log_4 x}$$

Therefore

$$4\log_4 x = 9\log_x 4$$

$$\Leftrightarrow 4\log_4 x = 9 \times \frac{1}{\log_4 x}$$

> We want to have logarithms involving just one base so that we can apply the rules of logarithms. Here we use the change of base rule to turn logs with base x into logs with base 4. (Alternatively, we could have turned them all into base x instead.)

$$\Leftrightarrow 4\left(\log_4 x\right)^2 = 9$$

$$\Leftrightarrow \left(\log_4 x\right)^2 = \frac{9}{4}$$

> Multiply through by $\log_4 x$ to get the log terms together.

> ⚠ Make sure you use brackets to indicate that the whole of $\log_4 x$ is being squared, not just x; $(\log_4 x)^2$ is **not** equal to $2\log_4 x$, but $\log_4 x^2$ would be.

$$\log_4 x = \frac{3}{2} \quad \text{or} \quad \log_4 x = -\frac{3}{2}$$

> Take the square root of both sides; don't forget the negative square root.

$$\text{So } x = 4^{\frac{3}{2}} \quad \text{or} \quad x = 4^{-\frac{3}{2}}$$

> Use $\log_a b = x \Leftrightarrow b = a^x$ to 'undo' the logs.

$$= 8 \qquad\qquad = \frac{1}{8}$$

Practice questions 1.4

 12. Solve the equation $\log_4 x + \log_4(x-6) = 2$.

 13. Solve the equation $2\log_2 x - \log_2(x+1) = 3$, giving your answers in simplified surd form.

> Make sure you check your answers by substituting them into the original equation.

 14. Solve the equation $25\log_2 x = \log_x 2$.

 15. Solve the equation $\log_4(4-x) = \log_{16}(9x^2 - 10x + 1)$.

1.5 PROBLEMS INVOLVING EXPONENTIAL FUNCTIONS

WORKED EXAMPLE 1.5

When a cup of tea is made, its temperature is 85°C. After 3 minutes the tea has cooled to 60°C. Given that the temperature T (°C) of the cup of tea decays exponentially according to the function $T = A + Ce^{-0.2t}$, where t is the time measured in minutes, find:

(a) the values of A and C (correct to three significant figures)

(b) the time it takes for the tea to cool to 40°C.

(a) When $t = 0$: $85 = A + C$ $\quad \cdots (1)$

When $t = 3$: $60 = A + Ce^{-0.6}$ $\cdots (2)$

$(1) - (2)$ gives $25 = C(1 - e^{-0.6})$

So $C = \dfrac{25}{1 - e^{-0.6}} = 55.4$ (3 SF)

> Substitute the given values for T (temperature) and t (time) into $T = A + Ce^{-0.2t}$, remembering that $e^0 = 1$.

Then, from (1):

$A = 85 - C = 85 - 55.4 = 29.6$ (3 SF)

> Note that A is the long-term limit of the temperature, which can be interpreted as the temperature of the room.

(b) When $T = 40$:

$29.6 + 55.4e^{-0.2t} = 40$

> Now we can substitute for T, A and C.

$\Rightarrow e^{-0.2t} = \dfrac{40 - 29.6}{55.4}$

$\Rightarrow \ln(e^{-0.2t}) = \ln\left(\dfrac{40 - 29.6}{55.4}\right)$

$\Rightarrow -0.2t = \ln\left(\dfrac{40 - 29.6}{55.4}\right)$

$\Rightarrow t = 8.36$ minutes

> Since the unknown t is in the power, we take logarithms of both sides and then 'cancel' e and ln using $\log_a(a^x) = x$.

 Remember that ln means $\log_e$.

Practice questions 1.5

16. The amount of reactant, V (grams), in a chemical reaction decays exponentially according to the function $V = M + Ce^{-0.32t}$, where t is the time in seconds since the start of the reaction. Initially there was 4.5 g of reactant, and this had decayed to 2.6 g after 7 seconds.

(a) Find the value of C.

(b) Find the value that the amount of reactant approaches in the long term.

17. A population of bacteria grows according to the model $P = Ae^{kt}$, where P is the size of the population after t minutes. Given that after 2 minutes there are 200 bacteria and after 5 minutes there are 1500 bacteria, find the size of the population after 10 minutes.

Mixed practice 1

1. Solve the equation $3 \times 9^x - 10 \times 3^x + 3 = 0$.

2. Find the exact solution of the equation $2^{3x+1} = 5^{5-x}$.

3. Solve the simultaneous equations
$$\ln x^2 + \ln y = 15$$
$$\ln x + \ln y^3 = 10$$

4. Given that $y = \ln x - \ln(x + 2) + \ln(x^2 - 4)$, express x in terms of y.

 5. The graph with equation $y = 4\ln(x - a)$ passes through the point $(5, \ln 16)$. Find the value of a.

6. (a) An economic model predicts that the demand, D, for a new product will grow according to the equation $D = A - Ce^{-0.2t}$, where t is the number of days since the product launch. After 10 days the demand is 15 000 and it is increasing at a rate of 325 per day.

   (i)   Find the value of C.

   (ii) Find the initial demand for the product.

   (iii) Find the long-term demand predicted by this model.

   (b) An alternative model is proposed, in which the demand grows according to the formula $D = B\ln\left(\dfrac{t+10}{5}\right)$. The initial demand is the same as that for the first model.

   (i)   Find the value of B.

   (ii) What is the long-term prediction of this model?

   (c) After how many days will the demand predicted by the second model become larger than the demand predicted by the first model?

Going for the top 1

1. Find the exact solution of the equation $2^{3x-4} \times 3^{2x-5} = 36^{x-2}$, giving your answer in the form $\dfrac{\ln p}{\ln q}$ where p and q are integers.

2. Given that $\log_a b^2 = c^2$ and $\log_b a = c + 1$, express a in terms of b.

3. In a physics experiment, Maya measured how the force, F, exerted by a spring depends on its extension, x. She then plotted the values of $a = \ln F$ and $b = \ln x$ on a graph, with b on the horizontal axis and a on the vertical axis. The graph was a straight line, passing through the points $(2, 4.5)$ and $(4, 7.2)$. Find an expression for F in terms of x.

2 POLYNOMIALS

WHAT YOU NEED TO KNOW

- The quadratic equation $ax^2 + bx + c = 0$ has solutions given by the quadratic formula:

$$x = \frac{-b \pm \sqrt{b^2 - 4ac}}{2a}$$

- The number of real solutions to a quadratic equation is determined by the discriminant, $\Delta = b^2 - 4ac$.

  - If $\Delta > 0$, there are two distinct solutions.

  - If $\Delta = 0$, there is one (repeated) solution.

  - If $\Delta < 0$, there are no real solutions.

- The graph of $y = ax^2 + bx + c$ has a y-intercept at $(0, c)$ and a line of symmetry at $x = -\dfrac{b}{2a}$.

- The graph of $y = a(x - p)(x - q)$ has x-intercepts at $(p, 0)$ and $(q, 0)$.

- The graph of $y = a(x - h)^2 + k$ has a turning point at (h, k).

- An expression of the form $(a + b)^n$ can be expanded quickly using the binomial theorem:

$$(a + b)^n = a^n + \binom{n}{1}a^{n-1}b + \ldots + \binom{n}{r}a^{n-r}b^r + \ldots + b^n$$

- The binomial coefficients can be found using a calculator, Pascal's triangle or the formula

$$\binom{n}{r} = \frac{n!}{r!(n-r)!}$$

⚠ EXAM TIPS AND COMMON ERRORS

- Make sure that you rearrange quadratic equations so that one side is zero before using the quadratic formula.

- Questions involving the discriminant are often disguised. You may have to interpret them to realise that you need to find the *number* of solutions rather than the *actual* solutions.

- Look out for quadratic expressions in disguise. A substitution is often a good way of making the expression explicitly quadratic.

2.1 USING THE QUADRATIC FORMULA

WORKED EXAMPLE 2.1

 Solve the equation $x^2 = 4x + 3$, giving your answer in the form $a \pm \sqrt{b}$.

$x^2 - 4x - 3 = 0$
Here $a = 1, b = -4$ and $c = -3$

> Rearrange the equation to make one side zero; then use the quadratic formula.

$$x = \frac{-(-4) \pm \sqrt{(-4)^2 - 4 \times 1 \times (-3)}}{2 \times 1}$$

$$= \frac{4 \pm \sqrt{28}}{2}$$

$$= \frac{4 \pm \sqrt{4} \times \sqrt{7}}{2}$$

$$= \frac{4 \pm 2\sqrt{7}}{2}$$

> Use the fact that $\sqrt{ab} = \sqrt{a} \times \sqrt{b}$ to simplify the answer.

$$= 2 \pm \sqrt{7}$$

Practice questions 2.1

1. Solve the equation $12x = x^2 + 34$, giving your answer in the form $a \pm \sqrt{b}$.

2. Find the exact solutions of the equation $x + \dfrac{1}{x} = 4$.

> ⚠ An exact solution in this context means writing your answer as a surd. Even giving all the decimal places shown on your calculator is not 'exact'.

3. Solve the equation $x^2 + 8k^2 = 6kx$, giving your answer in terms of k.

4. Using the substitution $u = x^2$, solve the equation $x^4 - 5x^2 + 4 = 0$.

5. A field is 6 m wider than it is long. The area of the field is 50 m². Find the exact dimensions of the field.

2.2 USING THE DISCRIMINANT

WORKED EXAMPLE 2.2

The line $y = kx - 1$ is tangent to the curve $y = x^2$. Find the possible values of k.

 This type of question can be done using calculus (by finding the equation of the tangent), but using the discriminant makes it much easier.

▶▶ Using calculus to find the equations of tangents (and normals) is covered in Chapter 7.

$y = kx - 1$ ··· (1)

$y = x^2$ ··· (2)

Substituting (2) *into* (1):

$x^2 = kx - 1$

$\Rightarrow x^2 - kx + 1 = 0$

We use the fact that the tangent and the curve intersect at only one point to deduce that there can only be one solution to their simultaneous equations.

Since there can only be one root, $b^2 - 4ac = 0$:

$(-k)^2 - 4 \times 1 \times 1 = 0$

$\Leftrightarrow k^2 = 4$

$\Leftrightarrow k = \pm 2$

For this quadratic to have only one solution, the discriminant must be zero.

 Do not forget the plus or minus symbol. The question asks for value**s** rather than a value, indicating that there is more than one solution.

Practice questions 2.2

6. Find the set of values of k for which the equation $2x^2 - x + k = 0$ has equal roots.

 These are all ways of giving information about the *number* of roots of a quadratic equation. Make sure you are able to interpret them in this context.

7. Find the values of k such that the quadratic expression $kx^2 + 12x + 6$ is always positive.

8. Given that the line $y = ax - 5$ is a tangent to the curve $y = 3x^2 + x - 2$, find the possible values of a.

2.3 THE COMPLETED SQUARE FORM $y = a(x - h)^2 + k$

WORKED EXAMPLE 2.3

(a) Write the expression $f(x) = 2x^2 + 12x + 10$ in the form $a(x + h)^2 + k$.

(b) Hence write down the coordinates of the vertex of the graph $y = 2x^2 + 12x + 10$.

(a) $a(x+h)^2 + k = ax^2 + 2ahx + ah^2 + k$

> Expand the brackets in the completed square form $a(x + h)^2 + k$.

$a = 2$, $2ah = 12$ and $ah^2 + k = 10$

So $h = 3$ and $k = -8$

Therefore $2x^2 + 12x + 10 = 2(x+3)^2 - 8$

> Compare coefficients with the given expression. (Comparing coefficients is one of several ways of completing the square.)

(b) The vertex is $(-3, -8)$.

> We must be careful to get the signs correct here. Comparing with the standard completed square form $a(x - h)^2 + k$, we see that the x-coordinate of the minimum point is −3.

 The command 'hence write down' suggests that the result can be taken immediately from the previous answer.

Practice questions 2.3

9. (a) Write the expression $x^2 - 9x + 4$ in the form $(x - h)^2 + k$.

(b) Hence write down the coordinates of the turning point on the graph $y = x^2 - 9x + 4$.

10. (a) Write the expression $3x^2 + 18x + 20$ in the form $a(x + h)^2 + k$.

(b) Hence find the exact solution of the equation $3x^2 + 18x + 20 = 0$.

11. (a) Write the expression $x^2 + bx + c$ in the form $(x + h)^2 + k$.

(b) Hence show that the solution of the equation $x^2 + bx + c = 0$ is $x = -\dfrac{b}{2} \pm \sqrt{\dfrac{b^2}{4} - c}$.

12. Explain why the minimum point of the curve $y = (x - h)^2 + k$ is (h, k).

2.4 GETTING QUADRATIC FUNCTIONS FROM THEIR GRAPHS

WORKED EXAMPLE 2.4

The diagram shows the graph of the quadratic function $f(x) = ax^2 + bx + c$.

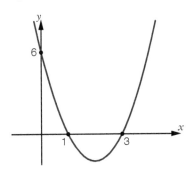

Find the values of a, b and c.

Since the roots are 1 and 3, the function can be written as $a(x-1)(x-3)$.

> Since we are given the x-intercepts, $f(x) = a(x - p)(x - q)$ is the most useful form of a quadratic function to use.

When $x = 0, f(x) = 6$, so
$a \times (-1) \times (-3) = 6$
$a = 2$

> When only one unknown parameter is left, we can use a third known point on the graph to find the value of the parameter.

$2(x-1)(x-3) = 2(x^2 - 4x + 3) = 2x^2 - 8x + 6$

So $a = 2, b = -8$ and $c = 6$.

> Expand the brackets to put the expression into the correct form.

Practice questions 2.4

13. Find the equation of the graph of the quadratic function shown on the right.

14. Find the equation of a quadratic graph with x-intercepts 4 and −2 and y-intercept 6.

15. The quadratic graph $y = x^2 + bx + c$ has a minimum point at (4, 6). Find the values of b and c.

16. Find the equation of the quadratic graph which has an x-intercept at (4, 0) and vertex at (6, 3).

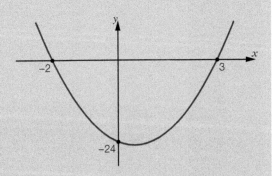

17. (a) The graph $y = x^2 + bx + c$ has a line of symmetry at $x = 5$ and a y-intercept at $y = 24$. Find the values of b and c.

    (b) Find the x-intercepts of this graph.

2.5 APPLYING THE BINOMIAL THEOREM

WORKED EXAMPLE 2.5

Find the coefficient of x^4 in the expansion of $\left(x - \dfrac{5}{x}\right)^6$.

The general term in the expansion is $\dbinom{6}{r}x^{6-r}\left(-\dfrac{5}{x}\right)^r$

 Start with the form of a general term. We could also have chosen $\dbinom{6}{r}x^6\left(-\dfrac{5}{x}\right)^{6-r}$, but this is algebraically harder to work with.

$$= \binom{6}{r}x^{6-r} \times x^{-r} \times (-5)^r$$

$$= \binom{6}{r}x^{6-2r} \times (-5)^r$$

Simplify using the rules of exponents.

⚠ Be careful with negative signs in the binomial expansion. Ensure they stay within brackets so that the power is applied to the negative sign.

Require that $6 - 2r = 4$

$\therefore 2r = 2$

and so $r = 1$

We need the term in x^4, so equate that to the power of x in the general term.

The relevant term is therefore

$$\binom{6}{1}x^4(-5)^1 = -30x^4$$

 Substitute $r = 1$ into $\dbinom{6}{r}x^{6-2r} \times (-5)^r$ to find the required term.

So the coefficient is -30.

⚠ Make sure you answer the question; you were asked to find the coefficient and not the whole term.

Practice questions 2.5

18. Find the constant term in the expansion of $\left(x^2 - \dfrac{3}{x}\right)^9$.

19. The third term in the binomial expansion of $(1 + px)^7$ is $84x^2$. Find:
    (a) the possible values of p
    (b) the fifth term in the expansion.

Mixed practice 2

1. Find the exact solutions of the equation $x^2 + 8x + 3 = 0$, simplifying your answers as far as possible.

2. The graph of a quadratic function passes through the points $(a - 4, 0)$ and $(a + 2, 0)$, and the coordinates of its vertex are $(5, -2)$. Find the value of a.

3. (a) Find the first four terms in ascending powers of x in the expansion of $(2 - x)^5$.

   (b) Hence find the value of 1.99^5 correct to 5 decimal places.

4. The curve $y = x^2 + kx + 2$ touches the x-axis. Find the possible values of k.

5. (a) Write $x^2 - 8x + 25$ in the form $(x - p)^2 + q$.

   (b) Hence find the maximum possible value of $\dfrac{3}{x^2 - 8x + 25}$.

6. Find the term that is independent of x in the expansion of $\left(x^3 + \dfrac{3}{x} \right)^8$.

7. Find the set of values of k for which the equation $3x - 2x^2 = k$ has no solutions.

8. Solve the equation $x^4 + x^2 - 20 = 0$.

9. (a) Factorise $ax - x^2$ and hence find the solutions to the equation $ax - x^2 = 0$.

   The graph shows the path of a football after Melissa kicks it. The path can be modelled by the curve $y = ax - x^2$, where y is the height of the ball in metres above the ground and x is the horizontal distance in metres travelled by the ball.

   (b) State the value of a.

   (c) Find the maximum height reached by the football.

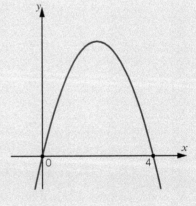

10. A graph has equation $y = x^2 + 4x - 12$.

    (a) Find the coordinates of the points where the line with equation $y = 6x - 9$ intersects the graph.

    (b) Find the coordinates of the points where the graph crosses the x-axis.

    A second graph has equation $y = ax^2 + bx + c$ and the same x-intercepts as the first graph, as shown below.

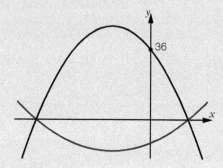

    (c) Find the values of a, b and c.

    (d) Find the distance between the vertices of the two graphs.

Going for the top 2

1. Find the quadratic term in the expansion of $(2 + x)(3 - 2x)^5$.

2. The graph of a quadratic function is shown in the diagram.

Find the equation of the graph.

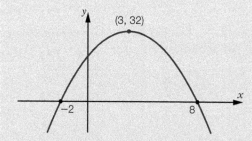

3. The coefficient of x^2 in the expansion of $(1 + ax)^n$ is 54 and the coefficient of x is 12. Find the values of a and n.

4. The graph with equation $y = kx^2 + 9x + k$ crosses the x-axis twice. Find the set of possible values of k.

3 FUNCTIONS, GRAPHS AND EQUATIONS

WHAT YOU NEED TO KNOW

- A function is a rule where for every input there is only one output. Graphically, any vertical line will cross the graph of a function no more than once.

  - The domain of a function is the set of allowed inputs. Typical reasons to restrict a domain include:

    - division by zero

    - square rooting of a negative number

    - taking the logarithm of a negative number or zero.

  - The range is the set of all possible outputs of a function. The easiest way of finding the range is to sketch the graph.

- A GDC can be used to identify the important features of a graph and to solve equations.

  - The y-intercept can be found by substituting $x = 0$ into the equation.

  - The x-intercepts are also called zeros or roots; they can be found by setting $y = 0$.

  - The maximum and minimum points need to be considered when finding the range.

   Maximum and minimum points can also be found by using differentiation. If the question does not explicitly ask you to use differentiation, you can just read the coordinates of maximum and minimum points off the graph.

  - The solutions of the equation $f(x) = g(x)$ are the intersection points of the graphs of $y = f(x)$ and $y = g(x)$.

- Composing functions means applying one function to the result of another:

  - $fg(x)$ is the function f applied to the output of the function g. This can also be written as $f \circ g(x)$ or $f(g(x))$.

- An inverse function, $f^{-1}(x)$, is a function which undoes the effect of another function, f.

  - $ff^{-1}(x) = f^{-1}f(x) = x$

  - To find $f^{-1}(x)$, write $y = f(x)$, rearrange to get x in terms of y, and then replace each y with an x.

  - Inverse functions have certain properties:

    - The graph of $y = f^{-1}(x)$ is the reflection of the graph $y = f(x)$ in the line $y = x$.

    - The domain of $f^{-1}(x)$ is the same as the range of $f(x)$. The range of $f^{-1}(x)$ is the same as the domain of $f(x)$.

- The rational function $f(x) = \dfrac{ax + b}{cx + d}$ has a graph called a hyperbola.

  - The x-intercept is at $x = -\dfrac{b}{a}$.

  - The y-intercept is at $y = \dfrac{b}{d}$.

  - The vertical asymptote is $x = -\dfrac{d}{c}$.

  - The horizontal asymptote is $y = \dfrac{a}{c}$.

- A change to a function results in a change to the graph of the function.

Transformation of $y = f(x)$	Transformation of the graph
$y = f(x) + c$	Translation $\begin{pmatrix} 0 \\ c \end{pmatrix}$
$y = f(x + d)$	Translation $\begin{pmatrix} -d \\ 0 \end{pmatrix}$
$y = pf(x)$	Vertical stretch of factor p, away from the x-axis for $p > 0$
$y = f(qx)$	Horizontal stretch of factor $\dfrac{1}{q}$, towards the y-axis for $q > 0$
$y = -f(x)$	Reflection in the x-axis
$y = f(-x)$	Reflection in the y-axis

⚠ EXAM TIPS AND COMMON ERRORS

- The range of a function is the same as the domain of the inverse function, which is sometimes a useful way of finding or checking the range.

- Some equations can only be solved approximately, using a GDC. On the calculator paper, do not spend too long trying to solve an equation using algebra. Always sketch the graph and mark the intersection points.

3.1 DOMAIN AND RANGE

WORKED EXAMPLE 3.1

If $f(x) = \dfrac{1}{\ln x}$:

(a) Find the largest possible domain of $f(x)$.

(b) Find the range of $f(x)$ if the domain is $x \geq e$.

(a) $x > 0$, as this is the domain of $\ln x$.

But also require $\ln x \neq 0$ to avoid division by zero, so $x \neq 1$.

Hence the domain is $x > 0, x \neq 1$.

 The two restrictions relevant here are logs having to take a positive argument and not dividing by zero.

(b)

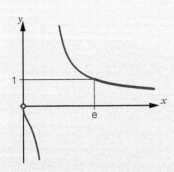

⚠ Always sketch the graph when finding the range.

The range is $0 < y \leq 1$.

 Use the graph to state what y values can occur. It is not obvious from the GDC that the x-axis is an asymptote here. You need to know that $\ln x$ slowly gets larger as x gets larger, which means that $y = \dfrac{1}{\ln x}$ slowly tends to zero, i.e. the x-axis.

Practice questions 3.1

 1. Find the largest possible domain of the function $f(x) = \dfrac{1}{\sqrt{9 - x^2}}$.

 2. Find the domain of the function $f(x) = \dfrac{1}{1-x} - \dfrac{1}{1+x}$.

⚠ If a question asks for 'the domain' of a function, you should give **the largest possible real domain**.

3. Find in terms of a the range of the function $y = x^2 - 6ax + a^2$.

3.2 INVERSE AND COMPOSITE FUNCTIONS

WORKED EXAMPLE 3.2

If $f(x) = \dfrac{1}{\sqrt{x}+1}$, $x \geq 0$ and $g(x) = 3x - 1$, solve the equation $gf^{-1}(x) = 11$.

$y = \dfrac{1}{\sqrt{x}+1} \Rightarrow \sqrt{x}+1 = \dfrac{1}{y}$

$\Rightarrow \sqrt{x} = \dfrac{1}{y} - 1$

$\Rightarrow x = \left(\dfrac{1}{y} - 1\right)^2$

> We first need to find $f^{-1}(x)$.
> Set $y = f(x)$, rearrange to make x the subject, and then replace each y with an x.

The domain of $f^{-1}(x)$ is the range of $f(x)$, which is $0 < y \leq 1$.

Therefore $f^{-1}(x) = \left(\dfrac{1}{x} - 1\right)^2$ for $0 < x \leq 1$.

$gf^{-1}(x) = g\left(\left(\dfrac{1}{x} - 1\right)^2\right) = 3\left(\dfrac{1}{x} - 1\right)^2 - 1$

for $0 < x \leq 1$.

> Compose the two functions by replacing each x in $g(x)$ with $\left(\dfrac{1}{x} - 1\right)^2$.

$3\left(\dfrac{1}{x} - 1\right)^2 - 1 = 11$

$\Leftrightarrow \left(\dfrac{1}{x} - 1\right)^2 = 4$

$\Leftrightarrow \dfrac{1}{x} - 1 = \pm 2$

> Here it is easier to take the square root of both sides than to multiply out the brackets, but remember the + and – signs when doing so.

$\therefore x = \dfrac{1}{3}$ or $x = -1$

But $0 < x \leq 1$.

Practice questions 3.2

4. If $f(x) = e^{ax+b}$, find $f^{-1}(x)$.

5. If $f(x) = \dfrac{x}{1 - \sqrt{x}}$, $x \geq 0$ and $g(x) = 3x + 1$, solve the equation $f^{-1}g(x) = \dfrac{9}{16}$.

Hint: do not attempt to find $f^{-1}(x)$.

6. A function is defined in the following table:

x	1	2	3	4	5	6	7	8	9
$f(x)$	7	1	6	4	2	4	9	8	3

   (a) Find $f \circ f(3)$.
   (b) Find $f^{-1}(9)$.

3.3 TRANSFORMATIONS OF GRAPHS

WORKED EXAMPLE 3.3

If $f(x) = x^3 + \sin x$:

(a) Find the resulting function when the graph of $f(x)$ is transformed by applying a translation with vector $\begin{pmatrix} 3 \\ 6 \end{pmatrix}$ followed by a vertical stretch with scale factor 2 away from the x-axis.

(b) Describe the transformation which transforms the graph of $y = f(x)$ to the graph of $g(x) = 8x^3 + \sin(2x)$.

(a) Translation with vector $\begin{pmatrix} 3 \\ 6 \end{pmatrix}$:

$$f(x-3)+6 = (x-3)^3 + \sin(x-3)+6$$

Then, applying a vertical stretch with scale factor 2:

$$2(f(x-3)+6) = 2\left((x-3)^3 + \sin(x-3)+6\right)$$

(b) $8x^3 + \sin(2x) = (2x)^3 + \sin(2x) = f(2x)$

This is a horizontal stretch away from the y-axis with scale factor $\dfrac{1}{2}$.

○— Relate each transformation to function notation:
$f(x - 3)$ translates the graph 3 units to the right;
$f(x) + 6$ translates the graph 6 units up.

○— $2f(x)$ stretches the graph vertically with scale factor 2.

⚠ You don't need to simplify the final expression unless a particular form is specified.

○— Write $g(x)$ in terms of $f(x)$ and relate this form to a transformation.

Practice questions 3.3

7. Sketch the graph of $y = -\sin 2x$ for $0 \le x \le \pi$.

8. Describe two transformations which transform the graph of $y = x^2$ to the graph of:
   (a) $y = 3(x-2)^2$
   (b) $y = 3x^2 - 2$

9. The diagram shows the graph of $y = f(x)$.
   Sketch the graph of $y = 3f(2x)$.

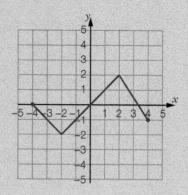

3.4 RATIONAL FUNCTIONS

WORKED EXAMPLE 3.4

Sketch the graph of $y = \dfrac{x+1}{2-x}$, labelling all the axis intercepts and asymptotes.

(a) When $y = 0$, $x = -1$; when $x = 0$, $y = \dfrac{1}{2}$.

Asymptotes: vertical $x = 2$, horizontal $y = -1$.

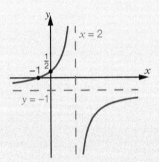

> This is a rational function of the form
> $f(x) = \dfrac{ax+b}{cx+d}$, so the graph is a hyperbola.
> We can find the x- and y-intercepts by substituting $y = 0$ and $x = 0$, respectively.
> The vertical asymptote occurs when the denominator is zero and the horizontal asymptote is at $y = \dfrac{a}{c}$.

Practice questions 3.4

10. Sketch the graph of $y = \dfrac{2x - a}{x + b}$ where a and b are positive constants. State the equations of all the asymptotes and the coordinates of axis intercepts.

11. If $f(x) = \dfrac{3x - 1}{2x + 3}$:

    (a) Find an expression for $f^{-1}(x)$.

    (b) State the domain and range of

        (i) $f(x)$ (ii) $f^{-1}(x)$

12. The diagram shows the graph of $y = f(x)$.

    (a) On a separate diagram sketch the graph of $y = f^{-1}(x)$.

    (b) Given that $f(x) = \dfrac{ax - b}{x}$, find the values of a and b.

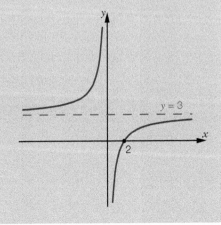

3.5 USING THE GDC TO SOLVE EQUATIONS

WORKED EXAMPLE 3.5

Solve the equation $3^{-x} = 4 - x^2$.

Using GDC to find the intersection of $y = 3^{-x}$ and $y = 4 - x^2$:

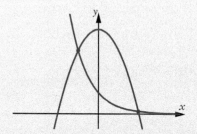

$x = -1$ or 1.97 (3 SF)

 There is no need to rearrange or simplify the equation in any way; simply sketch the graphs of the functions on both sides.

 Consider whether there might be further intersection points outside your viewing window.

The question is asking us to solve for x, so we just need to write down the x-coordinate of the intersection points.

Practice questions 3.5

 Equations can usually be solved in more than one way using the GDC. Try to select the most appropriate method for each equation.

13. Solve the equation $5^x = x^3 + 2$.

14. Find all the solutions of the equation $x - \dfrac{1}{x} = 4 - x^2$.

15. The first term of a geometric series is 2 and the sum of the first four terms is 17. Find the value of the common ratio.

▶▶ Geometric series are covered in Chapter 4.

 You can use the polynomial equation solver on your GDC, but the equation would first need to be rearranged into the form 'polynomial expression = 0'.

3.6 SOLVING PAIRS OF LINEAR EQUATIONS

WORKED EXAMPLE 3.6

A function is given by $g(x) = ax^3 - bx + 2$. Given that $g(-1) = 4$ and $g(2) = 16$, find the values of a and b.

$g(-1) = a(-1)^3 - b(-1) + 2 = 4 \Rightarrow -a + b = 2$

$g(2) = a(2)^3 - b(2) + 2 = 16 \Rightarrow 8a - 2b = 14$

 Find $g(-1)$ and $g(2)$ to form a pair of linear equations.

From GDC: $a = 3, b = 5$

 To solve a pair of linear equations using a graph, both equations must be put in the form $y = mx + c$. To solve using an equation solver, the correct form is $ax + by = c$.

Practice questions 3.6

16. The graph of the function $f(x) = A \times 3^x + k$ passes through the points $(1, 16)$ and $(2, 46)$. Find the values of A and k.

17. Find the coordinates of the point of intersection of the lines with equations $3x - 4y = 7$ and $11x + 5y = 74$.

18. The fifth term of an arithmetic sequence is 25 and the eighteenth term is 64. Find the first term and the common difference of the sequence.

> ▶ Arithmetic sequences are covered in Chapter 4.

19. A function $f(x) = ax^4 + bx + 1$ satisfies $f(2) = 42$ and $f'(2) = 21$. Find the values of a and b.

> ▶ Differentiation is covered in Chapter 7.

20. A bag of rice costs r dollars and a loaf of bread costs b dollars. Samir orders 26 bags of rice and 42 loaves of bread for his restaurant, and pays 272 dollars.

(a) Write an equation to represent the information given.

Anita buys 2 bags of rice and 7 loaves of bread for her party. She spends 26 dollars.

(b) Find the price of a loaf of bread.

Mixed practice 3

1. (a) Sketch the graph of $y = x^2 - x - 6$.

   (b) Hence state the domain of the function $f(x) = \ln(x^2 - x - 6)$.

2. Find, in terms of a, the domain of the function $f(x) = \sqrt{x^2 - a^2}$.

3. Below is the graph of $y = a\cos(bx + c)$. Find the values of a, b and c.

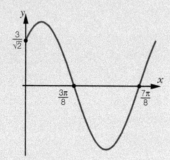

4. (a) Draw the curve $y = x + \dfrac{4}{x}$.

   (b) Find the exact solutions of the equation $x + \dfrac{4}{x} = 10$.

   (c) The equation $x + \dfrac{4}{x} = k$ has one solution. Find the two possible values of k.

5. The diagram shows the graph of the function $y = f(x)$. On separate diagrams, sketch the graphs of:

   (a) $y = 3f(x + 2)$

   (b) $y = f(-2x)$

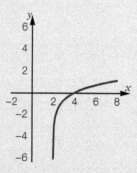

6. Butter is taken out of the fridge and left to warm up. The temperature of the butter, $T°\text{C}$, after m minutes is given by the equation $T = c - k \times 1.2^{-m}$. The initial temperature of the butter is $5°\text{C}$, and after 5 minutes its temperature has increased to $15°\text{C}$, as shown on the graph.

   (a) Find the values of k and c.

   (b) How long does it take for the butter to warm up to $20°\text{C}$?

   (c) Find the equation of the asymptote of the graph. What does this value represent?

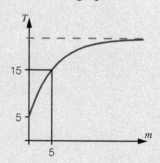

7. A function is defined by $g(x) = x^3 - ax + b$. The graph of $y = g(x)$ has gradient 3 at the point $(2, -2)$. Find the values of a and b.

8. If $f(x) = \dfrac{2x - 1}{x + 2}$:

   (a) Sketch the graph of $y = f(x)$.

   (b) Find an expression for $f^{-1}(x)$ and state its range.

   (c) Find the domain of the function $g(x) = f^{-1}(2 - x)$.

   (d) Solve the equation $f(x) = g(x)$.

Going for the top 3

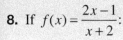

1. (a) On the same axes, sketch the graphs of $y = \ln x$ and $y = 2\ln(x - 2)$.

   (b) Solve the equation $\ln x = 2\ln(x - 2)$.

2. The graph of $y = 2x^3 - 5x$ is translated 3 units in the positive vertical direction and 2 units to the left, and is then reflected in the x-axis. Find the equation of the resulting graph in the form $y = ax^3 + bx^2 + cx + d$.

3. Consider a function, f, defined by $f(x) = \dfrac{ax - a^2 + 1}{x - a}$.

   (a) Show that $f(x) = p + \dfrac{q}{x - a}$, where p and q are constants to be determined in terms of a.

   (b) Hence give two transformations which change the graph of $y = \dfrac{1}{x}$ into the graph of $y = f(x)$.

   (c) Sketch the graph of $y = f(x)$, labelling all asymptotes and points of intersection with the axes.

   (d) State the domain and range of $f(x)$.

   (e) Find and simplify $f \circ f(x)$.

   (f) Find $f^{-1}(x)$.

   (g) What does your answer to part (f) tell you about the graph of $f(x)$?

4 SEQUENCES AND SERIES

- The notation for sequences and series:
  - u_n represents the nth term of the sequence u
  - S_n denotes the sum of the first n terms of the sequence
  - $\sum_{r=k}^{n} u_r$ denotes the sum of the kth term to the nth term, so $S_n = \sum_{r=1}^{n} u_r$
- A sequence can be described in two ways:
  - using a recursive definition to define how u_{n+1} depends on u_n
  - using a deductive rule (the nth term formula) to define how u_n depends on n
- An arithmetic sequence has a constant difference, d, between terms: $u_{n+1} = u_n + d$
  - $u_n = u_1 + (n-1)d$
  - $S_n = \dfrac{n}{2}(2u_1 + (n-1)d) = \dfrac{n}{2}(u_1 + u_n)$
- A geometric sequence has a constant ratio, r, between terms: $u_{n+1} = ru_n$
  - $u_n = u_1 r^{n-1}$
  - $S_n = \dfrac{u_1(r^n - 1)}{r-1} = \dfrac{u_1(1-r^n)}{1-r}, \; r \neq 1$
  - When $|r| < 1$, S_n approaches a limit as n increases, called the sum to infinity: $S_\infty = \dfrac{u_1}{1-r}$

⚠ EXAM TIPS AND COMMON ERRORS

- Many questions on sequences and series involve forming and then solving simultaneous equations.

- For questions on geometric sequences, you may need to use logarithms or rules of exponents.

- You may need to use the list feature on your calculator to solve problems involving sequences.

- You only ever need to use the first sum formula for geometric sequences.

- Geometric sequences are often used to solve problems involving percentage increase or decrease, such as investments and mortgages.

4.1 ARITHMETIC SEQUENCES AND SERIES

WORKED EXAMPLE 4.1

The third term of an arithmetic sequence is 15 and the sixth term is 27.

(a) Find the tenth term.

(b) Find the sum of the first ten terms.

(c) The sum of the first n terms is 5250. Find the value of n.

 You will often have to find a term of a sequence and a sum of a sequence in different parts of the same question. Make sure you are clear which you are being asked for.

(a) $u_3 = u_1 + 2d = 15$ ... (1)

$u_6 = u_1 + 5d = 27$ ... (2)

(2)−(1): $3d = 12 \Rightarrow d = 4$ and so $u_1 = 7$

Hence $u_{10} = u_1 + 9d = 7 + 9 \times 4 = 43$

> We can form two equations from the given information about the third and sixth terms.
>
> It is then clear that we need to solve these equations simultaneously for u_1 and d.

(b) $S_{10} = \dfrac{10}{2}(u_1 + u_{10})$

$= 5(7 + 43)$

$= 250$

> There are two possible formulae for the sum of an arithmetic sequence. Since we know the first term and the last term (from part (a)), we use
> $S_n = \dfrac{n}{2}(u_1 + u_n)$.

(c) $5250 = \dfrac{n}{2}(2u_1 + (n-1)d)$

$= \dfrac{n}{2}(14 + 4(n-1))$

The solutions (from GDC) are $n = 50$ or -52.5; but n must be a positive integer, so $n = 50$.

> Again we need a formula for the sum, but this time the other formula is the one to use as n is the only unknown here. Form an equation and solve it using a GDC.

Practice questions 4.1

1. The fifth term of an arithmetic sequence is 64 and the eighth term is 46.
   (a) Find the thirtieth term.
   (b) Find the sum of the first twelve terms.

2. The first four terms of an arithmetic sequence are 16, 15.5, 15, 14.5.
   (a) Find the twentieth term.
   (b) Which term is equal to zero?
   (c) The sum of the first n terms is 246. Find the possible values of n.

4.2 GEOMETRIC SEQUENCES AND SERIES

WORKED EXAMPLE 4.2

The second term of a geometric sequence is -5 and the sum to infinity is 12.
Find the common ratio and the first term.

$u_2 = u_1 r = -5 \quad \cdots (1)$

$S_\infty = \dfrac{u_1}{1-r} = 12 \quad \cdots (2)$

From (2): $u_1 = 12(1-r)$

Substituting into (1): $12r(1-r) = -5$

$\therefore r = -0.316$ or $r = 1.32$ (from GDC)

We can form two equations from the given information about the second term and sum to infinity.

It is then clear that we need to solve these simultaneously for u_1 and r.

 There is no need to rearrange the final equation as it can be solved using a GDC.

Since the sum to infinity exists, $|r| < 1$ and so $r = -0.316$.

Hence $u_1 = 12(1-(-0.316)) = 15.8$

The condition that $|r| < 1$ should be remembered as part of the formula for S_∞.

Practice questions 4.2

3. The fourth term of a geometric sequence is -16 and the sum to infinity is 32.
   Show that there is only one possible value of the common ratio and find this value.

4. The fifth term of a geometric series is 12 and the seventh term is 3.
   Find the two possible values of the sum to infinity of the series.

5. The sum of the first three terms of a geometric sequence is 38, and the sum of the first four terms is 65. Find the first term and the common ratio, $r > 1$.

6. The fifth term of a geometric sequence is 128 and the sixth term is 512.
   (a) Find the common ratio and the first term.
   (b) Which term has a value of 32 768?
   (c) How many terms are needed before the sum of all the terms in the sequence exceeds 100 000?

7. The first three terms of a geometric sequence are $2x + 4$, $x + 5$, $x + 1$, where x is a real number.
   (a) Find the two possible values of x.
   (b) Given that it exists, find the sum to infinity of the series.

4.3 APPLICATIONS

WORKED EXAMPLE 4.3

Daniel invests \$500 at the beginning of each year in a scheme that earns interest at a rate of 4% per annum, paid at the end of the year.

Show that the first year, n, in which the scheme is worth more than \$26 000 satisfies $n > \dfrac{\ln k}{\ln 1.04}$, where k is a constant to be found. Hence determine n.

Amount in the scheme at the end of the first year:
$$500 \times 1.04$$

Amount at the end of the second year:
$$(500 + 500 \times 1.04) \times 1.04$$
$$= 500 \times 1.04 + 500 \times 1.04^2$$

> Generate the first and second terms of the sequence to establish a pattern.

 With more complicated questions of this type, it is always a good idea to write down the first few terms to see whether you have an arithmetic or geometric series, and to understand exactly how the series is being formed.

So, amount at the end of the nth year is:
$$500 \times 1.04 + 500 \times 1.04^2 + \ldots + 500 \times 1.04^n$$

> This is a geometric series with $u_1 = 500 \times 1.04$ and $r = 1.04$.

$$S_n = \frac{500 \times 1.04(1.04^n - 1)}{1.04 - 1}$$
$$= 13000(1.04^n - 1)$$

So, for the amount to exceed \$26 000:

$$13000(1.04^n - 1) > 26000$$
$$\Rightarrow 1.04^n - 1 > 2$$
$$\Rightarrow 1.04^n > 3$$

> We can use the formula for the sum of the first n terms of a geometric sequence to form an inequality, which we then solve to find n.

$$\Rightarrow \ln(1.04^n) > \ln 3$$
$$\Rightarrow n \ln 1.04 > \ln 3$$
$$\Rightarrow n > \frac{\ln 3}{\ln 1.04} = 28.01 \,(2\,\text{DP})$$

Therefore, $k = 3$ and $n = 29$.

> The unknown n is in the power, so use logarithms to solve the inequality.

 Logarithms are covered in Chapter 1.

Practice questions 4.3

8. A starting salary for a teacher is $25 000 and there is an annual increase of 3%.
   (a) How much will the teacher earn in their 10th year?
   (b) How much will the teacher earn in total during a 35-year teaching career?
   (c) Find the first year in which the teacher earns more than $35 000.
   (d) How many years would the teacher have to work in order to earn a total of $1 million?

9. Beth repays a loan of $10 500 over a period of n months. She repays $50 in the first month, $55 in the second, and so on, with the monthly repayments continuing to increase by $5 each month.
   (a) How much will Beth repay in the 28th month?
   (b) Show that $n^2 + 19n - 4200 = 0$.
   (c) Hence find the number of months taken to repay the loan in full.

10. A ball is dropped from a height of 3 m. Each time it hits the ground it bounces up to 90% of its previous height.
    (a) How high does it reach on the fifth bounce?
    (b) On which bounce does the ball first reach a maximum height of less than 1 m?
    (c) Assuming the ball keeps bouncing until it rests, find the total distance travelled by the ball.

11. Theo has a mortgage of $127 000 which is to be repaid in annual instalments of $7000. Once the annual payment has been made, 3.5% interest is added to the remaining balance at the beginning of the next year.
    (a) Show that at the end of the third year the amount owing is given by
    $$127\,000 \times 1.035^2 - 7000\,(1 + 1.035 + 1.035^2)$$
    (b) By forming a similar expression for the amount owing after n years, show that
    $$n > \frac{\ln k}{\ln 1.035} + 1$$
    where k is a constant to be found.
    (c) Hence find the number of years it will take Theo to pay off his mortgage.

Mixed practice 4

1. The fourth term of an arithmetic sequence is 17. The sum of the first twenty terms is 990. Find the first term, a, and the common difference, d, of the sequence.

2. The fourth, tenth and thirteenth terms of a geometric sequence form an arithmetic sequence. Given that the geometric sequence has a sum to infinity, find its common ratio correct to three significant figures.

3. Evaluate $\sum_{r=1}^{12} 4r + \left(\dfrac{1}{3}\right)^r$ correct to four significant figures.

4. Find an expression for the sum of the first 20 terms of the series
   $$\ln x + \ln x^4 + \ln x^7 + \ln x^{10} + \dots$$

   giving your answer as a single logarithm.

5. A rope of length 300 m is cut into several pieces, whose lengths form an arithmetic sequence with common difference d. If the shortest piece is 1 m long and the longest piece is 19 m, find d.

6. Aaron and Blake each open a savings account. Aaron deposits $100 in the first month and then increases his deposits by $10 each month. Blake deposits $50 in the first month and then increases his deposits by 5% each month. After how many months will Blake have more money in his account than Aaron?

Going for the top 4

1. (a) (i) Prove that the sum of the first n terms of a geometric sequence with first term a and common ratio r is given by:
   $$S_n = \frac{a(r^n - 1)}{r - 1}$$

   (ii) Hence establish the formula for the sum to infinity, clearly justifying any conditions imposed on the common ratio r.

   (b) Show that in a geometric sequence with common ratio r, the ratio of the sum of the first n terms to the sum of the next n terms is $1 : r^n$.

   (c) In a geometric sequence, the sum of the seventh term and four times the fifth term equals the eighth term.

   (i) Find the ratio of the sum of the first 10 terms to the sum of the next 10 terms.

   (ii) Does the sequence have a sum to infinity? Explain your answer.

2. Find the sum of all integers between 1 and 1000 which are not divisible by 7.

5 TRIGONOMETRY

- The graphs of trigonometric functions:

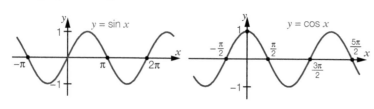 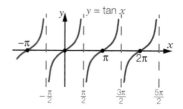

- The following exact values:

	0	$\dfrac{\pi}{6}$	$\dfrac{\pi}{4}$	$\dfrac{\pi}{3}$	$\dfrac{\pi}{2}$
sin	0	$\dfrac{1}{2}$	$\dfrac{\sqrt{2}}{2}$	$\dfrac{\sqrt{3}}{2}$	1
cos	1	$\dfrac{\sqrt{3}}{2}$	$\dfrac{\sqrt{2}}{2}$	$\dfrac{1}{2}$	0
tan	0	$\dfrac{1}{\sqrt{3}}$	1	$\sqrt{3}$	

- Trigonometric functions are related through identities:

  - Definition of tan:
    $$\tan\theta = \frac{\sin\theta}{\cos\theta}$$

  - Pythagorean identity:
    $$\cos^2\theta + \sin^2\theta = 1$$

  - Double angle identities:

    - $\sin 2\theta = 2\sin\theta\cos\theta$

    - $\cos 2\theta = \cos^2\theta - \sin^2\theta = 2\cos^2\theta - 1 = 1 - 2\sin^2\theta$

- When solving trigonometric equations of the form $\sin A = k$, $\cos A = k$ or $\tan A = k$:

  - Draw a graph to see how many solutions there are.

  - Use $\sin^{-1}$, $\cos^{-1}$ or $\tan^{-1}$ to find one possible value of A: A_0.

  - Find the second solution, A_1, by using the symmetry of the graph:

    - For $\sin A = k$, $A_1 = \pi - A_0$

    - For $\cos A = k$, $A_1 = 2\pi - A_0$

    - For $\tan A = k$, $A_1 = A_0 + \pi$

  - Find all the solutions in the required interval for A by adding multiples of 2π.

- With more complicated equations, it may be necessary to first use one or more of the identities to manipulate the equation into a form that can be solved.

- The sine and cosine rules are used to find the sides and angles of any triangle:

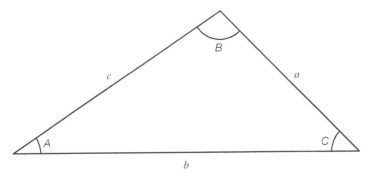

- Sine rule: $\dfrac{a}{\sin A} = \dfrac{b}{\sin B} = \dfrac{c}{\sin C}$

- Cosine rule: $c^2 = a^2 + b^2 - 2ab\cos C$, or $\cos C = \dfrac{a^2 + b^2 - c^2}{2ab}$

  - The sine rule is only used when a side and its opposite angle are given.

- The area of a triangle can be found using $\text{Area} = \dfrac{1}{2}ab\sin C$.

- When solving problems in three dimensions, look for right-angled triangles.

- In a circle of radius r with an angle of θ radians subtended at the centre:

  - the length of the arc AB: $l = r\theta$

  - the area of the sector AOB: $A = \dfrac{1}{2}r^2\theta$

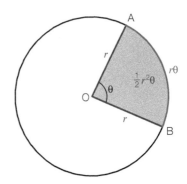

5.1 TRANSFORMATIONS OF TRIGONOMETRIC GRAPHS

WORKED EXAMPLE 5.1

The graph shown has equation $y = c - a\sin\left(\dfrac{x}{b}\right)$. Find the values of a, b and c.

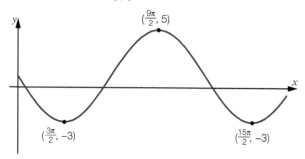

$y_{max} - y_{min} = 5 - (-3) = 8$

$\therefore a = \dfrac{8}{2} = 4$

> a has the effect of stretching the sine curve vertically. The distance between the maximum and minimum values of the original sine curve is 2. The minus sign in front of a causes a vertical reflection; it does not affect the amplitude.

◄◄ Transforming graphs is covered in Chapter 3.

$\text{period} = \dfrac{15\pi}{2} - \dfrac{3\pi}{2} = 6\pi$

$\therefore b = \dfrac{6\pi}{2\pi} = 3$

> Dividing x by b stretches the sine curve horizontally by a factor of b. The period of the original sine curve is 2π.

$c = 1$

> c causes a vertical translation. After multiplication by $-a = -4$, the minimum and maximum values would be -4 and 4 respectively, but here they are -3 and 5.

Practice questions 5.1

1. The graph shown has equation $y = a\sin\left(bx - \dfrac{\pi}{6}\right)$.
Find the values of a and b.

2. Find the range of the function $f(x) = \dfrac{2}{5 + 2\sin x}$.

3. Find the smallest positive value of x for
which $\dfrac{1}{3 - \sin\left(x - \dfrac{\pi}{4}\right)}$ takes its maximum value.

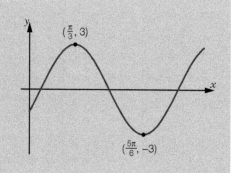

5.2 USING IDENTITIES TO FIND EXACT VALUES OF TRIGONOMETRIC FUNCTIONS

WORKED EXAMPLE 5.2

Given that $\cos A = \dfrac{1}{3}$ and $A \in \left[0, \dfrac{\pi}{2}\right]$, find the exact values of:

(a) $\tan A$ (b) $\sin 2A$

(a) $\sin^2 A = 1 - \cos^2 A$

$$= 1 - \frac{1}{9} = \frac{8}{9}$$

$$\therefore \sin A = \pm\sqrt{\frac{8}{9}} = \pm\frac{2\sqrt{2}}{3}$$

> $\tan A = \dfrac{\sin A}{\cos A}$, so we need to find $\sin A$ from the given value of $\cos A$. To do this we can use the identity $\sin^2 A + \cos^2 A = 1$.

> ⚠ Remember the $\pm$ when you take the square root.

For $A \in \left[0, \dfrac{\pi}{2}\right]$, $\sin A \ge 0$

$$\therefore \sin A = \frac{2\sqrt{2}}{3}$$

$$\Rightarrow \tan A = \frac{\sin A}{\cos A} = \frac{\frac{2\sqrt{2}}{3}}{\frac{1}{3}} = 2\sqrt{2}$$

> We now have to choose either the positive or the negative value, so consider the sine function in the given interval $0 \le A \le \dfrac{\pi}{2}$.

(b) $\sin 2A = 2\sin A \cos A$

$$= 2\left(\frac{2\sqrt{2}}{3}\right)\left(\frac{1}{3}\right)$$

$$= \frac{4\sqrt{2}}{9}$$

> As we need $\sin 2A$, the only option is to use the sine double angle identity.
> Substitute the values for $\sin A$ and $\cos A$.

Practice questions 5.2

4. Given that $\cos 2\theta = \dfrac{2}{3}$ and $\theta \in \left[\dfrac{\pi}{2}, \pi\right]$, find the exact value of $\cos\theta$.

5. Given that $\cos\theta = \dfrac{3}{4}$ and $\theta \in \left[\dfrac{3\pi}{2}, 2\pi\right]$, find the exact value of $\sin\theta$.

6. Given that $\tan A = 2$ and $A \in \left[0, \dfrac{\pi}{2}\right]$, find the exact value of $\sin A$.

5.3 SOLVING TRIGONOMETRIC EQUATIONS

WORKED EXAMPLE 5.3

 Solve the equation $\sin 3x = \dfrac{1}{2}$ for $x \in \left[0, \dfrac{\pi}{2}\right]$.

$\sin^{-1}\left(\dfrac{1}{2}\right) = \dfrac{\pi}{6}$ — First find one possible value for $3x$.

$x \in \left[0, \dfrac{\pi}{2}\right] \Rightarrow 3x \in \left[0, \dfrac{3\pi}{2}\right]$ — Find the range of possible values for $3x$.

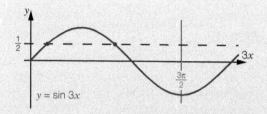
$y = \sin 3x$

Draw the graph to see how many solutions there are in the required interval. In this case there are only two.

$3x = \dfrac{\pi}{6}$ or $\pi - \dfrac{\pi}{6} = \dfrac{5\pi}{6}$

$\therefore x = \dfrac{\pi}{18}$ or $\dfrac{5\pi}{18}$

For the sine function, the second value will always be $\pi -$ first value.

Finally, divide by 3 to find x.

 It is a common error to divide by 3 first and then find the second value. Always find all the possible values in the interval first and then make x the subject at the end.

Practice questions 5.3

 7. Solve the equation $\dfrac{1}{\cos 2x} = \sqrt{2}$ for $x \in [0, \pi]$.

8. Solve the equation $\tan 3x = \sqrt{3}$ for $x \in [-\pi, \pi]$.

 9. Solve the equation $\sin\left(3x + \dfrac{\pi}{4}\right) = \dfrac{\sqrt{3}}{2}$ for $x \in [-\pi, \pi]$.

10. Solve the equation $8\sin^2 2x = 6$ for $0 \le x \le 2\pi$.

5.4 USING IDENTITIES TO SOLVE TRIGONOMETRIC EQUATIONS

WORKED EXAMPLE 5.4

Solve the equation $3\sin^2 x + 1 = 4\cos x$ for $x \in [-\pi, \pi]$.

$3\sin^2 x + 1 = 4\cos x$

$\Leftrightarrow 3(1 - \cos^2 x) + 1 = 4\cos x$

$\Leftrightarrow 3 - 3\cos^2 x + 1 = 4\cos x$

> Using the identity $\sin^2 x + \cos^2 x = 1$, we can replace the $\sin^2$ term and thereby form an equation that contains only one type of trigonometric function (cos).

 If possible, use an identity to ensure that there is only one type of trigonometric function in the equation.

$\Leftrightarrow 3\cos^2 x + 4\cos x - 4 = 0$

$\Leftrightarrow (3\cos x - 2)(\cos x + 2) = 0$

$\Leftrightarrow \cos x = \dfrac{2}{3}$ or $\cos x = -2$

$\therefore \cos x = \dfrac{2}{3}$ (as $-1 \leq \cos x \leq 1$)

$\cos^{-1}\left(\dfrac{2}{3}\right) = 0.841$ (3 SF)

> This now becomes a standard quadratic equation, which can be factorised and solved.

> ◄ Disguised quadratics are also encountered in Chapter 1 when solving exponential equations.

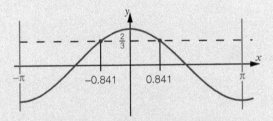

> Draw the graph to see how many solutions there are in the required interval. Here only two are needed.

$2\pi - 0.841 = 5.44$ is not in the interval

$5.44 - 2\pi$ is in the interval

$\therefore x = \pm 0.841$

> For the cosine function, the second value will always be 2π − first value, but here that is not in the required interval. Therefore, subtract 2π from this value to get the second answer.

Practice questions 5.4

11. Solve the equation $3\sin^2 \theta - 5\cos\theta = 1$ for $\theta \in [-\pi, \pi]$.

12. Solve the equation $\cos 2x - 2 = 5\cos x$ for $x \in [0, 2\pi]$.

 13. Solve the equation $2\sin x = \tan x$ for $-\pi \leq x \leq \pi$.

14. Find the values of $\theta \in [0, \pi]$ for which $\sin^2 2\theta - 4\cos 2\theta + 5\cos^2 2\theta = 0$.

5.5 GEOMETRY OF TRIANGLES AND CIRCLES

WORKED EXAMPLE 5.5

The diagram shows a circle with centre C and radius 12 cm. Points P and Q are on the circumference on the circle and $P\hat{C}Q = 0.72$ radians.

(a) Find the length of the chord PQ.

(b) Find the area of the shaded region.

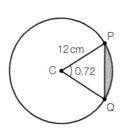

(a) Using the cosine rule in triangle PCQ:

$$PQ^2 = 12^2 + 12^2 - 2 \times 12 \times 12 \cos 0.72$$

$$= 71.5$$

$$\therefore PQ = 8.45 \, cm$$

CP = CQ = 12 since CP and CQ are both radii. As we don't have an angle and its opposite side, we need to use the cosine rule.

🖩 Make sure your calculator is in radian mode.

(b) Area of sector PCQ:

$$\frac{1}{2}r^2\theta = \frac{1}{2} \times 12^2 \times 0.72 = 51.84$$

Area of triangle PCQ:

$$\frac{1}{2}ab\sin C = \frac{1}{2}(12)(12)\sin 0.72 = 47.5$$

We can find the area of the shaded region by subtracting the area of the triangle from the area of the sector.

Shaded area = $51.84 - 47.5 = 4.36 \, cm^2$

Practice questions 5.5

15. A sector of a circle with angle 0.65 radians has area 14.8 cm². Find the radius of the circle.

16. The diagram shows a rectangle and a sector of a circle. Find the perimeter of the shape.

17. The sides of a triangle have lengths 12 cm, 17 cm and 13 cm. Find, in degrees, the size of the largest angle of the triangle.

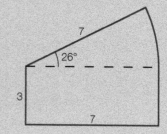

5.6 THE AMBIGUOUS CASE OF THE SINE RULE

WORKED EXAMPLE 5.6

In triangle ABC, AB = 10 cm, AC = 12 cm, and $A\hat{C}B = 46°$. Find the possible values of $A\hat{B}C$.

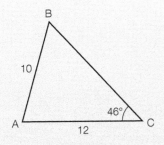

Using the sine rule:	Start by sketching a diagram.

$$\frac{\sin A\hat{B}C}{12} = \frac{\sin 46°}{10}$$

The sine rule can be used as we have a side and the angle opposite.

$$\Rightarrow \sin A\hat{B}C = \frac{12\sin 46°}{10} = 0.8632...$$

$\sin^{-1}(0.8632...) = 59.678...$
So $A\hat{B}C = 59.7°$ or $180° - 59.7° = 120°$ (3 SF)

Both θ and 180° − θ are possible answers (since sin θ = sin(180° − θ)).

 When using the sine rule, always check whether two solutions are possible. The question will usually alert you to look out for this.

If $A\hat{B}C = 59.7°$, the third angle will be
$180° - 59.7° - 46° = 74.3°$, which is fine.
If $A\hat{B}C = 120°$, the third angle will be
$180° - 120° - 46° = 14°$, which is also fine.

We need to check whether each solution is possible by finding the remaining angle using the fact that the three angles sum to 180°.
In this case, there are two possible triangles.

 If the second solution is impossible, the third angle will turn out to be negative.

Practice questions 5.6

18. In triangle ABC, AB = 6, AC = 14 and $A\hat{C}B = 20°$. Find the two possible lengths BC.

19. In triangle ABC, AB = 22 cm, AC = 14 cm and $A\hat{C}B = 58°$. Show that there is only one possible triangle with these measurements, and find its area.

Mixed practice 5

1. The area of the triangle shown in the diagram is $12\,\text{cm}^2$.
   Find the value of x.

   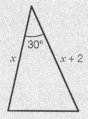

2. The triangle shown in the diagram has angles θ and 2θ. Find the value of θ in degrees.

   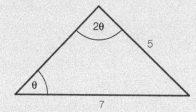
   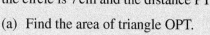

3. The depth of water in a harbour is modelled by the equation $d = 14 - 1.2\cos\left(\dfrac{\pi t}{12}\right)$, where d is measured in metres and t is the time in hours after midnight.

   (a) What is the first time after midnight at which the water depth is $14\,\text{m}$?

   (b) What is the smallest possible depth?

   (c) Find the times, within the first 24 hours, when the depth of water is less than $13.5\,\text{m}$.

4. Solve the equation $3\sin 2\theta = \tan 2\theta$ for $\theta \in \left[-\dfrac{\pi}{2}, \dfrac{\pi}{2} \right]$.

5. In the diagram, O is the centre of the circle and PT is the tangent to the circle at T. The radius of the circle is $7\,\text{cm}$ and the distance PT is $12\,\text{cm}$.

   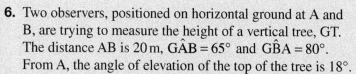

   (a) Find the area of triangle OPT.

   (b) Find the size of $\text{P}\hat{\text{O}}\text{T}$.

   (c) Find the area of the shaded region.

6. Two observers, positioned on horizontal ground at A and B, are trying to measure the height of a vertical tree, GT. The distance AB is $20\,\text{m}$, $\text{G}\hat{\text{A}}\text{B} = 65°$ and $\text{G}\hat{\text{B}}\text{A} = 80°$. From A, the angle of elevation of the top of the tree is $18°$.

   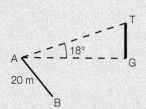

   (a) Find the distance between A and the bottom of the tree.

   (b) Find the height of the tree.

7. The area of the shaded region is $6.2\,\text{cm}^2$.

   (a) Show that $\theta - \sin\theta = 0.496$.

   (b) Find the value of θ in radians.

   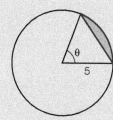

8. A straight line passes through the origin and the point (a, b), and makes an angle of θ with the positive x-axis.

  (a) Express b in terms of a and θ.

  (b) Hence find the gradient of the line.

  (c) Find the angle that the line with equation $y = 3x - 2$ makes with the positive x-axis.

9. A small Ferris wheel has radius $6\,\text{m}$, and the bottom of the wheel is $1\,\text{m}$ above the ground. A car is attached to one arm of the wheel, which makes an angle of $\theta°$ with the horizontal, as shown in the diagram.

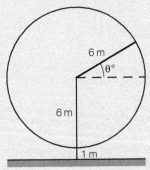

  (a) Find the height of the car above the ground when $\theta = 50$.

  (b) Find the values of θ for which the car is exactly $10\,\text{m}$ above the ground.

  (c) Assuming that the wheel rotates at a constant speed, for what proportion of time is the car more than $10\,\text{m}$ above the ground?

10. (a) The function f is defined by $f(x) = 3x^2 - 2x + 5$ for $-1 \le x \le 1$. Find the coordinates of the vertex of the graph of $y = f(x)$.

  (b) The function g is defined by $g(\theta) = 3\cos 2\theta - 4\cos\theta + 13$ for $0 \le \theta \le 2\pi$.

   (i) Show that $g(\theta) = 6\cos^2\theta - 4\cos\theta + 10$.

   (ii) Hence find the minimum value of $g(\theta)$.

11. In the diagram, AC is the diameter of the semicircle, BD is perpendicular to AC, AB = 2 and BC = 1, and $\hat{BAD} = \hat{CAE} = \theta$. Let $R = AD - CE$.

  (a) Find an expression for R in terms of θ.

  (b) Show that R has a stationary value when $2\sin\theta = 3\cos^3\theta$.

  (c) Assuming that this stationary value is a minimum, find the smallest possible value of AD − CE.

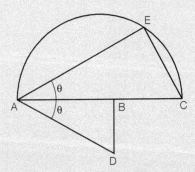

Going for the top 5

1. A vertical cliff, BT, of height 50 m stands on horizontal ground.
   The angle of depression of the top of a lighthouse, L, from the top of the cliff is 20°.
   The angle of elevation of L from the bottom of the cliff is 15°.
   Find the height of the lighthouse.

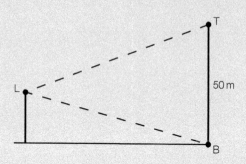

> ⚠ The angle of elevation is the angle above the horizontal. The angle of depression is the angle below the horizontal.

2. In triangle ABC, AB = 6 cm, AC = 10 cm, and $A\hat{C}B = 34°$. Show that there are two possible triangles with these measurements, and find the difference between their areas.

3. (a) Given that $\dfrac{1}{\tan x} - \tan x = 2$ and $x \in \left[0, \dfrac{\pi}{2}\right]$, find the exact value of $\tan x$.

   (b) (i) Show that $\dfrac{1}{\tan x} - \tan x = \dfrac{2}{\tan 2x}$.

   (ii) Hence solve the equation $\dfrac{1}{\tan x} - \tan x = 2$ for $x \in [0, \pi]$.

   (c) Find the exact value of $\tan\left(\dfrac{\pi}{8}\right)$.

6 VECTORS

WHAT YOU NEED TO KNOW

- A vector can represent the position of a point relative to the origin (position vector) or the displacement from one point to another.

  - The components of a position vector are the coordinates of the point.

  - If points A and B have position vectors a and b, the displacement vector $\overrightarrow{AB} = b - a$.

  - The position vector of the midpoint of [AB] is $\frac{1}{2}(a + b)$.

- A vector (in 3-dimensional space) can be represented by its components:

$$v = \begin{pmatrix} v_1 \\ v_2 \\ v_3 \end{pmatrix} = v_1 i + v_2 j + v_3 k$$

  - The magnitude of v is $|v| = \sqrt{v_1^2 + v_2^2 + v_3^2}$.

  - The unit vector in the same direction as v is $\hat{v} = \frac{1}{|v|} v$.

  - The distance between two points A and B, with position vectors a and b, is $AB = |b - a|$.

- The scalar product (or dot product) of two vectors $v = \begin{pmatrix} v_1 \\ v_2 \\ v_3 \end{pmatrix}$ and $w = \begin{pmatrix} w_1 \\ w_2 \\ w_3 \end{pmatrix}$ is given by

$$v \cdot w = v_1 w_1 + v_2 w_2 + v_3 w_3$$

  - The scalar product has many properties similar to those of multiplication:

    - $a \cdot b = b \cdot a$

    - $(ka) \cdot b = k(a \cdot b)$

    - $a \cdot (b + c) = a \cdot b + a \cdot c$

- The scalar product can be used to calculate the angle between two vectors:

$$\cos \theta = \frac{v \cdot w}{|v||w|}$$

  - If vectors a and b are perpendicular, $a \cdot b = 0$.

  - If vectors a and b are parallel, $a \cdot b = |a||b|$; in particular, $a \cdot a = |a|^2$.

    - For parallel vectors a and b, $a = \lambda b$ for some non-zero scalar λ.

- A straight line can be described by a vector equation
  $r = a + \lambda d$, where:

  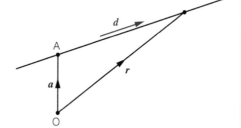

  - $d = \begin{pmatrix} d_1 \\ d_2 \\ d_3 \end{pmatrix}$ is the direction vector of the line

  - $A(a_1, a_2, a_3)$ is one point on the line.

    - Different values of the parameter λ give the
      positions of different points on the line.

- To find the point of intersection of two lines $r = a_1 + \lambda d_1$ and $r = a_2 + \mu d_2$, set the two position
  vectors equal to each other to form three separate equations. Solve two of these simultaneously
  to find the unknown parameters λ and μ, checking that the parameters also satisfy the third
  equation.

  - If two lines have parallel direction vectors, they are either parallel or coincident
    (the same line).

- To find the angle between two lines $r = a_1 + \lambda d_1$ and $r = a_2 + \mu d_2$, find the angle between their

  direction vectors using $\cos\theta = \dfrac{d_1 \cdot d_2}{|d_1||d_2|}$.

- An object starting at the point with position vector a and moving with (constant)
  velocity vector v moves along a path given by $r = a + tv$, where t is the time after the start
  of the motion.

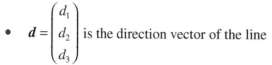 **EXAM TIPS AND COMMON ERRORS**

- Use a vector diagram to show all the given information and add to it as you work through a
  question.

- In long questions, it may be possible to answer a later part without having done all the
  previous parts.

- Be clear on the difference between the coordinates of a point, (a_1, a_2, a_3), and the position

  vector of the point, $\begin{pmatrix} a_1 \\ a_2 \\ a_3 \end{pmatrix}$.

6.1 PROVING GEOMETRICAL PROPERTIES USING VECTORS

WORKED EXAMPLE 6.1

ABCD is a square. Points M, N, P and Q lie on the sides [AB], [BC], [CD] and [DA] so that
$AM:MB = BN:NC = CP:PD = DQ:QA = 2:1$. Let $\overrightarrow{AB} = \boldsymbol{b}$ and $\overrightarrow{AD} = \boldsymbol{d}$.

Prove that [QN] and [MP] are perpendicular.

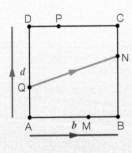

> ⚠ Always draw a diagram for questions involving geometrical figures.

$$\overrightarrow{QN} = \overrightarrow{QD} + \overrightarrow{DC} + \overrightarrow{CN}$$
$$= \frac{2}{3}\boldsymbol{d} + \boldsymbol{b} + \frac{1}{3}(-\boldsymbol{d})$$
$$= \boldsymbol{b} + \frac{1}{3}\boldsymbol{d}$$

> We need an expression for $\overrightarrow{QN}$ in terms of $\boldsymbol{b}$ and $\boldsymbol{d}$. Note that a ratio of 2:1 means $\frac{2}{3}$ and $\frac{1}{3}$, respectively, of the total length, and that moving backwards along a vector (in this case $\boldsymbol{d}$) means a negative sign.

$$\overrightarrow{MP} = \overrightarrow{MB} + \overrightarrow{BC} + \overrightarrow{CP}$$
$$= \frac{1}{3}\boldsymbol{b} + \boldsymbol{d} + \frac{2}{3}(-\boldsymbol{b}) = -\frac{1}{3}\boldsymbol{b} + \boldsymbol{d}$$

> Similarly, we find an expression for $\overrightarrow{MP}$ in terms of $\boldsymbol{b}$ and $\boldsymbol{d}$.

$$\overrightarrow{MP} \cdot \overrightarrow{QN} = \left(-\frac{1}{3}\boldsymbol{b} + \boldsymbol{d}\right) \cdot \left(\boldsymbol{b} + \frac{1}{3}\boldsymbol{d}\right)$$

> To prove that $\overrightarrow{MP}$ and $\overrightarrow{QN}$ are perpendicular, we need to show that $\overrightarrow{MP} \cdot \overrightarrow{QN} = 0$.

$$= -\frac{1}{3}\boldsymbol{b} \cdot \boldsymbol{b} + \boldsymbol{d} \cdot \boldsymbol{b} - \frac{1}{9}\boldsymbol{b} \cdot \boldsymbol{d} + \frac{1}{3}\boldsymbol{d} \cdot \boldsymbol{d}$$

> The brackets can be expanded with the scalar product just as with normal multiplication.

$$= -\frac{1}{3}|\boldsymbol{b}|^2 + \boldsymbol{b} \cdot \boldsymbol{d} - \frac{1}{9}\boldsymbol{b} \cdot \boldsymbol{d} + \frac{1}{3}|\boldsymbol{d}|^2$$

> Use the facts that $\boldsymbol{b} \cdot \boldsymbol{d} = \boldsymbol{d} \cdot \boldsymbol{b}$ and $\boldsymbol{b} \cdot \boldsymbol{b} = |\boldsymbol{b}|^2$.

$$= \frac{8}{9}\boldsymbol{b} \cdot \boldsymbol{d} = 0$$

> Since the sides of a square have equal length and are perpendicular, $|\boldsymbol{b}| = |\boldsymbol{d}|$ and $\boldsymbol{b} \cdot \boldsymbol{d} = 0$.

Hence [QN] and [MP] are perpendicular.

Practice questions 6.1

1. Points A, B, C and D have position vectors $\boldsymbol{a}, \boldsymbol{b}, \boldsymbol{c}$ and $\boldsymbol{d}$, respectively. M is the midpoint of [AB] and N is the midpoint of [BC].

 (a) Express $\overrightarrow{MN}$ in terms of $\boldsymbol{a}, \boldsymbol{b}$ and $\boldsymbol{c}$.

 P is the midpoint of [CD] and Q is the midpoint of [DA].

 (b) By finding an expression for $\overrightarrow{QP}$ in terms of $\boldsymbol{a}, \boldsymbol{c}$ and $\boldsymbol{d}$, show that MNPQ is a parallelogram.

6.2 EQUATION OF A LINE AND THE INTERSECTION OF LINES

WORKED EXAMPLE 6.2

(a) Find a vector equation of the line l_1 passing through points A(2, 1, 3) and B(-1, 1, 2).

(b) Show that l_1 does not intersect the line l_2 with equation $r = \begin{pmatrix} 1 \\ -1 \\ 0 \end{pmatrix} + t \begin{pmatrix} 2 \\ 1 \\ 3 \end{pmatrix}$.

(a) $r = a + \lambda d$

$$d = \overrightarrow{AB} = b - a = \begin{pmatrix} -1 \\ 1 \\ 2 \end{pmatrix} - \begin{pmatrix} 2 \\ 1 \\ 3 \end{pmatrix} = \begin{pmatrix} -3 \\ 0 \\ -1 \end{pmatrix}$$

We need the position vector of one point on the line and the direction vector of the line: a is the position vector of point A on the line, and $\overrightarrow{AB}$ is a vector in the direction of the line.

So a vector equation is $r = \begin{pmatrix} 2 \\ 1 \\ 3 \end{pmatrix} + \lambda \begin{pmatrix} -3 \\ 0 \\ -1 \end{pmatrix}$

⚠ We could also have used B as the point on the line and/or taken $\overrightarrow{BA}$ to be our d.

(b) For l_1:

$$r = \begin{pmatrix} 2-3\lambda \\ 1 \\ 3-\lambda \end{pmatrix}$$

and for l_2:

$$r = \begin{pmatrix} 1+2t \\ -1+t \\ 3t \end{pmatrix}$$

The two vectors on the RHS give expressions for the components of a position vector of a point on each line.

If the two lines intersect, there must be a value of t and a value of λ that give the same position vector.

If the lines pass through the same point:

$$\begin{cases} 2-3\lambda = 1+2t \\ 1 = -1+t \\ 3-\lambda = 3t \end{cases}$$

From the second equation, $t = 2$, and substituting this into the first gives $\lambda = -1$.

When two vectors are equal, all three components have to be equal. This gives three equations with two unknowns.

We can find t and λ from the first two equations, and we must then check whether these values also satisfy the third equation.

Checking the third equation:

$3-(-1) \neq 3 \times 2$

So the lines do not intersect.

This shows that there are no values of t and λ which make the two position vectors equal.

Practice questions 6.2

2. (a) Find a vector equation of the line passing through the points A(−3, 1, 2) and B(−3, 3, 7).

   (b) Find the coordinates of the point where this line meets the line with equation

$$r = \begin{pmatrix} 1 \\ -1 \\ -2 \end{pmatrix} + \lambda \begin{pmatrix} -2 \\ 3 \\ 7 \end{pmatrix}.$$

3. Determine whether the lines $r = \begin{pmatrix} 2 \\ 3 \\ 8 \end{pmatrix} + s \begin{pmatrix} 1 \\ 4 \\ 0 \end{pmatrix}$ and $r = \begin{pmatrix} 5 \\ 1 \\ 3 \end{pmatrix} + t \begin{pmatrix} -1 \\ 10 \\ 2 \end{pmatrix}$ intersect.

   If they do, find the point of intersection.

4. Two lines have equations $r = \begin{pmatrix} 5 \\ 1 \\ -3 \end{pmatrix} + \lambda \begin{pmatrix} 2 \\ -3 \\ 1 \end{pmatrix}$ and $r = \begin{pmatrix} 3 \\ a \\ b \end{pmatrix} + \mu \begin{pmatrix} 6 \\ p \\ q \end{pmatrix}.$

   (a) Find the values of p and q for which the two lines are parallel.

   (b) For those values of p and q, find the values of a and b for which the two lines are coincident.

5. The line m passes through the points A(−1, 3, 3) and B(7, 2, 2), and intersects the x-axis at the point Q.

   (a) Find the equation of line m.

   (b) Find the coordinates of point Q.

   (c) Calculate the distance of Q from A.

6. (a) Determine whether the point (−4, 1, 2) lies on the line with equation

$$r = \begin{pmatrix} 2 \\ -1 \\ 3 \end{pmatrix} + \lambda \begin{pmatrix} 2 \\ -1 \\ 1 \end{pmatrix}.$$

   The line l is parallel to the line above and passes through the point (1, 4, −3).

   (b) Write down the equation of line l.

   (c) Determine whether l intersects the z-axis.

6.3 ANGLES BETWEEN LINES

WORKED EXAMPLE 6.3

Find the acute angle between the lines with equations $r = (-i + 2j + 3k) + \lambda(2i + j - 2k)$ and $r = (2i + 3j + k) + \mu(i - j + 3k)$.

> ⚠ You can find the angle between two lines even if they do not intersect.

The direction vectors of the two lines are:

$d_1 = 2i + j - 2k$
$d_2 = i - j + 3k$

> We need to find the angle between the two direction vectors.

$$\cos\theta = \frac{d_1 \cdot d_2}{|d_1||d_2|}$$

> Use the scalar product formula.

$$= \frac{2 - 1 - 6}{\sqrt{4+1+4}\sqrt{1+1+9}}$$

$$= -0.5025\ldots$$

$$\therefore \theta = \cos^{-1}(-0.5025\ldots) = 120.2°.$$

So the acute angle is
$180° - 120.2° = 59.8°$

> We were asked for the *acute* angle.

Practice questions 6.3

7. Find the acute angle between the lines with equations $r = (i + 2k) + \lambda(3i + 4j + 12k)$ and $r = -2j + \mu(2i + 2j + k)$.

8. Find the acute angle that the line with equation $r = \begin{pmatrix} 0 \\ 4 \\ 0 \end{pmatrix} + \lambda \begin{pmatrix} 2 \\ -1 \\ 5 \end{pmatrix}$ makes with the y-axis.

> The y-axis points in the direction of the unit vector j.

9. The vertices of a triangle are the points A(−1, 2, 5), B(1, 1, 6) and C(−3, 5, 7).
   (a) Find the length of the side [BC].
   (b) Find the size of angle B in degrees.

6.4 APPLYING VECTORS TO MOTION

WORKED EXAMPLE 6.4

An aircraft taking off from $0i + 10j + 0k$ km moves with velocity $150i + 150j + 40k$ km h^{-1}, where the vector i represents East, j represents North and k represents vertically up.

(a) If t is the time in hours, write down a vector equation for the motion of the aircraft.

(b) Find the speed of the aircraft.

(c) Find the angle of elevation of the aircraft.

(a) $r = \begin{pmatrix} 0 \\ 10 \\ 0 \end{pmatrix} + t \begin{pmatrix} 150 \\ 150 \\ 40 \end{pmatrix}$

> We can use the standard form $r = a + tv$ to describe the path of the aircraft.

> ⚠ You can use either i, j, k notation or column vector notation; whichever you are more comfortable with, but do not forget that an equation needs '$r =$'.

(b) Speed $= \sqrt{150^2 + 150^2 + 40^2} \approx 216$ km h^{-1}

> ◯— Speed is the modulus of the velocity vector.

(c) The resultant speed of the components in the easterly and northerly directions is:
$\sqrt{150^2 + 150^2} \approx 212$ km h^{-1}
So the angle of elevation θ satisfies
$\tan\theta = \dfrac{40}{212}$
$\therefore \theta \approx 10.7°$

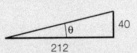

> ◯— The velocity is made up of a component 'along the ground' (the resultant of the i and j components) and a vertical (k) component.

> ⚠ Even though 212 km h^{-1} is written in the working, the calculation should be done using the full accuracy stored in the calculator.

Practice questions 6.4

10. The path of a flying bird is modelled by $r = 4ti + 6tj + (2 - t)k$ metres where t is in seconds.

  (a) Find the speed of the bird.

  (b) Find the angle of depression of its flight.

  (c) How far has the bird travelled when it lands on the ground?

11. A ship starts at $0i + 0j$ km and moves with velocity $20i + 5j$ km h^{-1}. A second ship starts at $7i + 12j$ km and moves with velocity $-19i - 8j$ km h^{-1}. Show that the ships have the same speed but do not collide.

Mixed practice 6

1. Point A has position vector $3i + 2j - k$. Point D lies on the line with equation $r = (i - j + 5k) + \lambda(-i + j - 2k)$. Find the value of λ such that (AD) is parallel to the x-axis.

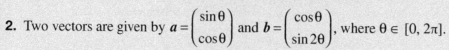

2. Two vectors are given by $a = \begin{pmatrix} \sin\theta \\ \cos\theta \end{pmatrix}$ and $b = \begin{pmatrix} \cos\theta \\ \sin 2\theta \end{pmatrix}$, where $\theta \in [0, 2\pi]$.

   Find all possible values of θ for which a and b are perpendicular.

3. (a) Find a vector equation of the line l passing through the points P(3, −1, 2) and Q(−1, 1, 7).

   (b) The point M has position vector $3i - 4j + k$. Find the acute angle between (PM) and l.

   (c) Hence find the shortest distance from M to the line l.

4. (a) Three points have coordinates A(3, 0, 2), B(−1, 4, 1) and C(−4, 1, 3). Find the coordinates of point D such that ABCD is a parallelogram.

   (b) Find the point of intersection of the diagonals of the parallelogram.

   (c) Find the acute angles between the diagonals.

   (d) State, with a reason, whether this parallelogram is a rectangle.

5. The angle between vectors $p = i + 2j + 2k$ and $q = xi + xj + 2k$ is 60°.

   (a) Find constants a, b and c such that $ax^2 + bx + c = 0$.

   (b) Hence find the angle between the vector q and the z-axis.

6. Three points have coordinates A(4, 1, 2), B(1, 5, 1) and C(λ, λ, 3).

   (a) Find the value of λ for which the triangle ABC has a right angle at B.

   (b) For this value of λ, find the coordinates of point D on the side [AC] such that AD = 2DC.

7. The line l_1 has vector equation $r = \lambda \begin{pmatrix} 3 \\ 5 \\ 1 \end{pmatrix}$ and the line l_2 has vector equation $r = \begin{pmatrix} 9 \\ 15 \\ 3 \end{pmatrix} + \mu \begin{pmatrix} 2 \\ 0 \\ -1 \end{pmatrix}$.

   (a) Show that the two lines meet and find the point of intersection.

   (b) Find the angle between the two lines.

   Two flies are flying in an empty room. One fly starts in a corner at position (0, 0, 0) and every second flies 3 cm in the x direction, 5 cm in the y direction and 1 cm up in the vertical z direction.

   (c) Find the speed of this fly.

   The second fly starts at the point (9, 15, 3) cm and each second travels 2 cm in the x direction and 1 cm down in the vertical z direction.

(d) Show that the two flies do not meet.

(e) Find the distance between the flies when they are at the same height.

(f) (i) Write down the displacement vector between the two flies at time t.

(ii) Show that at time t the distance, d, between the two flies satisfies
$d^2 = 315 - 180t + 30t^2$.

(iii) Hence find the minimum distance between the two flies.

Going for the top 6

1. Two lines have vector equations $l_1 : r = \begin{pmatrix} 1 \\ 3 \\ 1 \end{pmatrix} + \lambda \begin{pmatrix} 1 \\ -1 \\ 2 \end{pmatrix}$ and $l_2 : r = \begin{pmatrix} 5 \\ -1 \\ -6 \end{pmatrix} + \mu \begin{pmatrix} 1 \\ 1 \\ 3 \end{pmatrix}$.

Points A on l_1 and B on l_2 are such that [AB] is perpendicular to both lines.

(a) Show that $\mu_A - \lambda_B = 1$.

(b) Find another linear equation connecting λ_B and μ_A.

(c) Hence find the shortest distance between the two lines.

2. PQRS is a rhombus with $\overrightarrow{PQ} = a$ and $\overrightarrow{QR} = b$. The midpoints of the sides [PQ], [QR], [RS] and [SP] are A, B, C and D, respectively.

(a) Express $\overrightarrow{AB}$ and $\overrightarrow{BC}$ in terms of a and b.

(b) (i) Explain why $a \cdot a = b \cdot b$.

(ii) Show that [AB] and [BC] are perpendicular.

(c) What type of quadrilateral is ABCD?

3. (a) Show that the point A with coordinates $(-1, 5, 3)$ does not lie on the line l with equation
$$r = \begin{pmatrix} -2 \\ 3 \\ 1 \end{pmatrix} + \lambda \begin{pmatrix} 1 \\ 2 \\ -1 \end{pmatrix}.$$

(b) Find the coordinates of the point B on l such that [AB] is perpendicular to l.

7 DIFFERENTIATION

WHAT YOU NEED TO KNOW

- The derivative can be interpreted as the rate of change of one quantity as another changes, or as the gradient of a tangent to a graph.

- The notation for the derivative of $y = f(x)$ with respect to x is $\dfrac{dy}{dx}$ or $f'(x)$.

  - Differentiating again gives the second derivative, $\dfrac{d^2y}{dx^2}$ or $f''(x)$, which can be interpreted as the rate of change (or gradient) of $f'(x)$.

- Differentiation from first principles involves looking at the limit of the gradient of a chord. The formula is:

$$f'(x) = \lim_{h \to 0} \left(\frac{f(x+h) - f(x)}{h} \right)$$

- These derivatives are given in the Formula booklet:

$f(x)$	$f'(x)$
x^n	nx^{n-1}
$\sin x$	$\cos x$
$\cos x$	$-\sin x$
$\tan x$	$\dfrac{1}{\cos^2 x}$
e^x	e^x
$\ln x$	$\dfrac{1}{x}$

- The above derivatives can be combined by adding two together or multiplying by a constant:
  - $[f(x) + g(x)]' = f'(x) + g'(x)$
  - $[kf(x)]' = kf'(x)$

- The following derivatives of common functions composed with linear functions are very useful. (They are not in the Formula booklet but follow from those that are when the chain rule is applied.)

$f(x)$	$f'(x)$
$(ax+b)^n$	$an(ax+b)^{n-1}$
e^{ax+b}	ae^{ax+b}
$\ln(ax+b)$	$\dfrac{a}{ax+b}$
$\sin(ax+b)$	$a\cos(ax+b)$
$\cos(ax+b)$	$-a\sin(ax+b)$
$\tan(ax+b)$	$\dfrac{a}{\cos^2(ax+b)}$

- Further rules of differentiation:

  - The chain rule is used to differentiate composite functions:

    $$y = g(u) \text{ where } u = f(x) \implies \frac{dy}{dx} = \frac{dy}{du} \times \frac{du}{dx}$$

  - The product rule is used to differentiate two functions multiplied together:

    $$y = uv \implies \frac{dy}{dx} = u\frac{dv}{dx} + v\frac{du}{dx}$$

  - The quotient rule is used to differentiate one function divided by another:

    $$y = \frac{u}{v} \implies \frac{dy}{dx} = \frac{v\dfrac{du}{dx} - u\dfrac{dv}{dx}}{v^2}$$

- The equation of a tangent at the point (x_1, y_1) is given by $y - y_1 = m(x - x_1)$ where $m = f'(x_1)$.

  The equation of the normal at the same point is given by $y - y_1 = m(x - x_1)$ where $m = -\dfrac{1}{f'(x_1)}$.

- If a function is increasing, $f'(x) > 0$; if a function is decreasing, $f'(x) < 0$. If the graph is concave up, $f''(x) > 0$; if the graph is concave down, $f''(x) < 0$.

- Stationary points of a function are points where the gradient is zero, i.e. $f'(x) = 0$. The second derivative can be used to determine the nature of a stationary point.

  - At a local maximum, $f''(x) \leq 0$.

  - At a local minimum, $f''(x) \geq 0$.

  - At a point of inflexion, $f''(x) = 0$ but $f'''(x) \neq 0$; there is a change in concavity of the curve.

- Optimisation problems involve setting up an expression and then finding the maximum or minimum value by differentiating or using a GDC.
  - If there is a constraint, it will be necessary to set up a second expression from this information and then substitute it into the expression to be optimised, thereby eliminating the second variable.
  - The optimal solution may occur at an end point of the domain as well as at a stationary point.
- Differentiate with respect to time to change an expression for displacement (s) into an expression for velocity (v) and then into one for acceleration (a):
  - $v = \dfrac{ds}{dt}$
  - $a = \dfrac{dv}{dt} = \dfrac{d^2s}{dt^2}$

 EXAM TIPS AND COMMON ERRORS

- Be careful when the variable is in the power. If you differentiate e^x with respect to x the answer is e^x **not** xe^{x-1}.

- Always use the product rule to differentiate a product. You cannot simply differentiate each factor separately and multiply the answers together; the derivative of $x^2 \sin x$ is **not** $2x \cos x$.

- Make sure you are clear whether you have a product (such as $e^x \sin x$) or a composite function (such as $e^{\sin x}$) to differentiate. The latter is differentiated using the chain rule.

- It is sometimes easier to differentiate a quotient by turning it into a product (i.e. writing it as the numerator multiplied by the denominator raised to a negative power) and then differentiating using the product rule.

- Do not confuse the rules for differentiation and integration. Always check the sign when integrating or differentiating trigonometric functions, and carefully consider whether you should be multiplying or dividing by the coefficient of x.

- When differentiating trigonometric functions you **must** work in radians.

7.1 DIFFERENTIATION FROM FIRST PRINCIPLES

WORKED EXAMPLE 7.1

Use differentiation from first principles to prove that the derivative of x^3 is $3x^2$.

$f'(x) = \lim_{h \to 0} \left(\dfrac{f(x+h) - f(x)}{h} \right)$

$= \lim_{h \to 0} \left(\dfrac{(x+h)^3 - x^3}{h} \right)$

> We start with the definition of the derivative at the point x (i.e. the formula for differentiation from first principles).

$= \lim_{h \to 0} \left(\dfrac{x^3 + 3x^2h + 3xh^2 + h^3 - x^3}{h} \right)$

$= \lim_{h \to 0} \left(\dfrac{3x^2h + 3xh^2 + h^3}{h} \right)$

> We do not want to let the denominator tend to zero straight away, so first manipulate the numerator to get a factor of h that we can cancel with the h in the denominator.

$= \lim_{h \to 0} \left(3x^2 + 3xh + h^2 \right)$

> Divide top and bottom by h.

$= 3x^2$

> Once there is no h in the denominator we can let $h \to 0$.

Practice questions 7.1

1. Given that $y = 5x^2$, use differentiation from first principles to find $\dfrac{dy}{dx}$.

2. Prove from first principles that the derivative of x^4 is $4x^3$.

3. Prove from first principles that $\dfrac{d}{dx}(x^2 - 5x + 2) = 2x - 5$.

4. Let $f(x) = (x+1)(x-3)$.
   (a) Without expanding the brackets, use differentiation from first principles to find $f'(x)$.
   (b) Use the rules of differentiation to confirm that your expression for $f'(x)$ is correct.

7.2 THE PRODUCT, QUOTIENT AND CHAIN RULES

WORKED EXAMPLE 7.2

Differentiate $y = x \mathrm{e}^{\sin x}$.

Let $u = x$. Then $\dfrac{\mathrm{d}u}{\mathrm{d}x} = 1$

Let $v = \mathrm{e}^{\sin x}$

> This is a product, so we need to use the product rule. It doesn't matter which function is $u(x)$ and which is $v(x)$.

To find $\dfrac{\mathrm{d}v}{\mathrm{d}x}$, let $w = \sin x$. Then $v = \mathrm{e}^w$,

$\dfrac{\mathrm{d}w}{\mathrm{d}x} = \cos x$ and $\dfrac{\mathrm{d}v}{\mathrm{d}w} = \mathrm{e}^w$

> $v(x)$ is a composite function, so use the chain rule.

Therefore

$\dfrac{\mathrm{d}v}{\mathrm{d}x} = \dfrac{\mathrm{d}v}{\mathrm{d}w} \times \dfrac{\mathrm{d}w}{\mathrm{d}x}$

$\quad = \mathrm{e}^w \cos x$

$\quad = \mathrm{e}^{\sin x} \cos x$

> ⚠ You do not have to set out your working in this much detail; from $v = \mathrm{e}^{\sin x}$ you can proceed straight to $\dfrac{\mathrm{d}v}{\mathrm{d}x} = \mathrm{e}^{\sin x} \cos x$.

So $\dfrac{\mathrm{d}y}{\mathrm{d}x} = u\dfrac{\mathrm{d}v}{\mathrm{d}x} + v\dfrac{\mathrm{d}u}{\mathrm{d}x} = x\cos x \, \mathrm{e}^{\sin x} + \mathrm{e}^{\sin x}$

> Now apply the product rule.

> ⚠ There is no need to simplify (or factorise) your answer, unless you are asked to do so.

Practice questions 7.2

5. Differentiate $y = 4x^2 \cos 3x$.

 6. Find the gradient of the graph of $f(x) = \ln(3x^2 - 1)$ at the point where $x = 3$.

7. Differentiate $y = \mathrm{e}^{x^2} + \dfrac{\sin 3x}{2x}$.

8. Find the values of x for which the function $f(x) = \ln\left(\dfrac{2}{x^2 - 12}\right)$ has a gradient of 2.

9. Given that $f(x) = \dfrac{x^2 - 1}{x^2 + 2}$, find $f''(x)$ in the form $\dfrac{a - bx^2}{(x^2 + 2)^3}$.

7.3 TANGENTS AND NORMALS

WORKED EXAMPLE 7.3

Find the coordinates of the point where the normal to the curve $y = x^2$ at $x = a$ meets the curve again.

$$\frac{dy}{dx} = 2x$$

> The normal is perpendicular to the tangent, so we need the gradient of the tangent first.

At $x = a$, $\frac{dy}{dx} = 2a$, i.e. gradient of tangent is $2a$.

Therefore gradient of normal is $m = -\dfrac{1}{2a}$

Equation of normal: $y - a^2 = -\dfrac{1}{2a}(x - a)$

> The normal is a straight line, so its equation is of the form $y - y_1 = m(x - x_1)$.

Intersection with $y = x^2$:

$$x^2 - a^2 = -\frac{1}{2a}(x - a)$$

$$\Rightarrow (x - a)(x + a) = -\frac{1}{2a}(x - a)$$

> We need to find the point of intersection with $y = x^2$, so substitute $y = x^2$ into the equation of the normal.
>
> Factorise the left-hand side (LHS) so that we have a common factor on both sides.

$$\Rightarrow (x - a)\left(x + a + \frac{1}{2a}\right) = 0$$

So $x = a$ or $x = -a - \dfrac{1}{2a}$ ($x = a$ was given)

> Move everything to the LHS and factorise. Do not divide by $(x - a)$ as this could result in the loss of a solution. Instead, we find all possible solutions and then reject any that are not relevant.

When $x = -a - \dfrac{1}{2a}$, $y = \left(-a - \dfrac{1}{2a}\right)^2$

So the coordinates of the point are

$$\left(-a - \frac{1}{2a}, \left(-a - \frac{1}{2a}\right)^2\right)$$

> We can now find y by substituting into $y = x^2$.

> ⚠ If asked to find the coordinates of a point, make sure you find both x- and y-coordinates.

Practice questions 7.3

10. Find the equation of the normal to the curve $y = e^{-3x^2}$ at the point where $x = 2$.

11. A tangent to the curve $y = \tan x$ for $-\dfrac{\pi}{2} < x < \dfrac{\pi}{2}$ is drawn at the point where $x = \dfrac{\pi}{4}$. Find the x-coordinate of the point where this tangent intersects the curve again.

7.4 STATIONARY POINTS

WORKED EXAMPLE 7.4

Find and classify the stationary points on the curve $y = 2x^3 - 9x^2 + 12x + 5$.

$\dfrac{dy}{dx} = 6x^2 - 18x + 12$

For stationary points, $\dfrac{dy}{dx} = 0$:

$6x^2 - 18x + 12 = 0$

$\Leftrightarrow x^2 - 3x + 2 = 0$

$\Leftrightarrow (x-1)(x-2) = 0$

$\Leftrightarrow x = 1$ or $x = 2$

> At stationary points the first derivative is zero, so we need to find $\dfrac{dy}{dx}$ and then solve the equation $\dfrac{dy}{dx} = 0$.

When $x = 1, y = 2 - 9 + 12 + 5 = 10$

When $x = 2, y = 2(2)^3 - 9(2)^2 + 12(2) + 5 = 9$

So stationary points are $(1, 10)$ and $(2, 9)$.

> Find the y-coordinates and give the full coordinates of the stationary points.

$\dfrac{d^2y}{dx^2} = 12x - 18$

> We can use the second derivative to determine the nature of the stationary points.

 Make sure you differentiate the original expression for $\dfrac{dy}{dx}$ and not a manipulated version (such as $x^2 - 3x + 2$ here).

At $x = 1, \dfrac{d^2y}{dx^2} = 12 - 18 = -6 < 0$

$\therefore (1, 10)$ is a local maximum.

At $x = 2, \dfrac{d^2y}{dx^2} = 12(2) - 18 = 6 > 0$

$\therefore (2, 9)$ is a local minimum.

> Apply the second derivative test by substituting the x values into $\dfrac{d^2y}{dx^2}$.

Practice questions 7.4

12. Find and classify the stationary points on the curve $y = x^3 - 3x + 8$.

13. Find and classify the stationary points on the curve $y = x\sin x + \cos x$ for $0 < x < 2\pi$.

14. Find the maximum value of $y = \ln(x - \sin^2 x)$ for $0 < x \leq 2\pi$.

7.5 OPTIMISATION WITH CONSTRAINTS

WORKED EXAMPLE 7.5

What is the area of the largest rectangle that can just fit under the curve $y = \sin x$, $0 \le x \le \pi$, if one side of the rectangle lies on the x-axis?

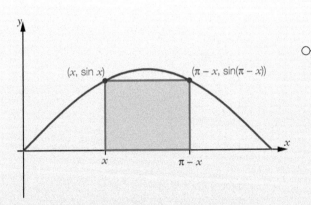

We start by defining the variables, taking the bottom left corner of the rectangle to be the point $(x, 0)$. Everything else then follows from the symmetry of the sine curve.

⚠️ With harder optimisation problems a major difficulty can be defining the variables, as there may be more than one choice. In an unfamiliar situation, it is often helpful to sketch a diagram.

Width of rectangle $= (\pi - x) - x = \pi - 2x$

So area $= (\pi - 2x)\sin x$

Find an expression for the quantity that needs to be optimised, in this case area.

Using a GDC for the graph of area against x:

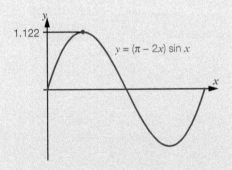

Since this is a calculator question, we do not need to differentiate; but we do need to sketch the graph to justify that we have found a maximum (rather than any other stationary point) and that the maximum is not at an end point.

Therefore, the maximum area is 1.122 (4 SF).

Practice questions 7.5

15. An open cylindrical can has radius r and height h. The height and the radius can both change, but the volume remains fixed at $64\pi \text{ cm}^2$. Find the minimum surface area of the can (including the base) and justify that the value you have found is a minimum.

16. Find the smallest surface area of a cone (including base) with volume 100 cm^3.

17. A square-based cuboid has surface area 520 cm^2. Prove that it takes the largest possible volume when it is a cube.

7.6 POINTS OF INFLEXION

WORKED EXAMPLE 7.6

The curve $y = x^3 - 6x^2 + 5x + 2$ has a point of inflexion. Find its coordinates.

$\dfrac{dy}{dx} = 3x^2 - 12x + 5$

$\Rightarrow \dfrac{d^2y}{dx^2} = 6x - 12$

At a point of inflexion the second derivative is zero, so we need to find $\dfrac{d^2y}{dx^2}$ and solve the equation $\dfrac{d^2y}{dx^2} = 0$.

At a point of inflexion, $\dfrac{d^2y}{dx^2} = 0$:

$6x - 12 = 0$

$\Leftrightarrow x = 2$

When $x = 2$:

$y = (2)^3 - 6(2)^2 + 5(2) + 2 = -4$

So the point of inflexion is at $(2, -4)$.

Find the y-coordinate.

 If a question states that a curve has a point of inflexion and there is only one solution to $\dfrac{d^2y}{dx^2} = 0$, you can assume that you have found the point of inflexion; there is no need to check that $\dfrac{d^3y}{dx^3} \neq 0$.

Practice questions 7.6

18. Find the coordinates of the point of inflexion on the curve $y = x^3 - 12x^2 + 7$.

19. The graph of $y = 4x^3 - ax^2 + b$ has a point of inflexion at $(-1, 4)$. Find the values of a and b.

20. A curve has equation $y = (x^2 - a)e^x$.
    (a) Show that at a point of inflexion, $x^2 + 4x - a + 2 = 0$.
    (b) Hence find the range of values of a for which the curve has at least one point of inflexion.
    (c) When $a = -1$, show that one of the points of inflexion is a stationary point.
    (d) For this value of a, sketch the graph of $y = (x^2 - a)e^x$.

7.7 INTERPRETING GRAPHS

WORKED EXAMPLE 7.7

The graph shows $y = f'(x)$.

On the graph:

(a) Mark points corresponding to a local maximum of $f(x)$ with an A.

(b) Mark points corresponding to a local minimum of $f(x)$ with a B.

(c) Mark points corresponding to a point of inflexion of $f(x)$ with a C.

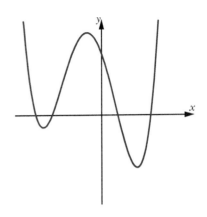

Local maximum points occur where the graph crosses the x-axis with a negative gradient.

 A local maximum has $f'(x) = 0$ and $f''(x) < 0$, i.e. the gradient of the graph of $y = f'(x)$ is negative there.

⚠ Be clear whether you are considering the graph of $y = f(x)$, $y = f'(x)$ or $y = f''(x)$.

Local minimum points occur where the graph crosses the x-axis with a positive gradient.

A local minimum has $f'(x) = 0$ and $f''(x) > 0$, i.e. the gradient of the graph of $y = f'(x)$ is positive there.

Points of inflexion occur at stationary points on the graph.

Therefore:

At a point of inflexion $f''(x) = 0$, i.e. the gradient of the graph of $y = f'(x)$ is zero (and the gradient is either positive on both sides of that point or negative on both sides).

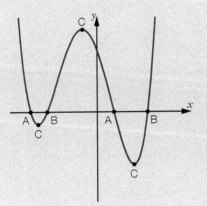

Practice questions 7.7

21. The graph shows $y = f'(x)$.

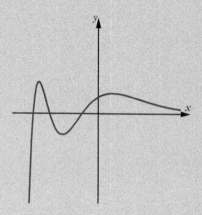

On the graph:
(a) Mark points corresponding to a local maximum of $f(x)$ with an A.
(b) Mark points corresponding to a local minimum of $f(x)$ with a B.
(c) Mark points corresponding to a point of inflexion of $f(x)$ with a C.
(d) Mark points corresponding to a zero of $f''(x)$ with a D.

22. The graph shows $y = f''(x)$.

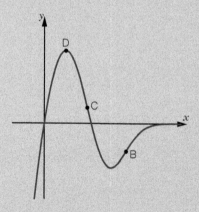

(a) On the graph, mark points corresponding to a point of inflexion of $f(x)$ with an A.
(b) State whether at the point B, the graph of $y = f(x)$ is concave up or concave down.
(c) Is $f'(x)$ increasing or decreasing at the point C?
(d) Given that the graph of $y = f(x)$ has a stationary point with the same x-coordinate as the point marked D, state the nature of this stationary point and justify your answer.

7.8 KINEMATICS

WORKED EXAMPLE 7.8

If the displacement, s, at time t is given by $s = 4t^2 - 3e^{-3t}$, find the time at which the minimum velocity occurs in the form $\ln k$ where k is a rational number.

$v = \dfrac{ds}{dt} = 8t + 9e^{-3t}$

> We first need to find an expression for the velocity.

$\dfrac{dv}{dt} = 8 - 27e^{-3t}$

At a local minimum, $\dfrac{dv}{dt} = 0$:

$8 - 27e^{-3t} = 0 \Rightarrow e^{3t} = \dfrac{27}{8}$

> If there is a local minimum, it will occur when $\dfrac{dv}{dt} = 0$.

$\Rightarrow t = \dfrac{1}{3}\ln\left(\dfrac{27}{8}\right) = \ln\left(\dfrac{3}{2}\right)$

> Use the laws of logarithms to put the result into the required form.

 Laws of logarithms are covered in Chapter 1.

$\dfrac{d^2v}{dt^2} = 81e^{-3t} > 0$ for all t

Therefore $t = \ln\left(\dfrac{3}{2}\right)$ is a minimum.

> We now need to check that this is a local minimum (rather than a maximum or point of inflexion). If it is not a minimum, the minimum velocity would occur at an end point.

Practice questions 7.8

23. A hiker has a displacement s km, at a time t hours, modelled by $s = t^3 - 4t$, $t \geq 0$.
    (a) Find the time it takes for the hiker to return to his original position (where he stops).
    (b) Find the maximum displacement from the starting point.
    (c) Find the maximum speed of the hiker.

24. A small ball oscillates on a spring so that its displacement from the starting position depending on time is given by $s = \dfrac{2}{3}\sin\left(\dfrac{3\pi t}{2}\right)$.
    (a) Find expressions for the velocity and acceleration of the ball at time t.
    (b) Find the first two values of t for which the speed of the ball equals $\dfrac{\pi}{2}$.

Mixed practice 7

1. Find the coordinates of the stationary point on the curve $y = 2x - e^x$.

2. A rectangle has perimeter $40\,\text{cm}$. One side of the rectangle has length $x\,\text{cm}$.

    (a) Find an expression for the area of the rectangle in terms of x.

    (b) Prove that the rectangle with the largest area is a square.

3. Find the equation of the normal to the curve $y = 5\sin 3x + x^2$ where $x = \pi$, giving your answer in the form $y = mx + c$.

4. A particle is moving with displacement s at time t.

    (a) A model of the form $s = at^2 + bt$ is applied. Show that the particle moves with constant acceleration.

    It is known that when $t = 1$, $v = 1$ and when $t = 2$, $v = 5$.

    (b) Find the values of a and b.

5. Find the coordinates of the points on the curve $y = 3\ln x + \dfrac{1}{x}$ where the tangent is parallel to the line $2x - y = 4$.

6. The cubic graph $y = ax^3 + bx^2 + cx + d$ has one stationary point. Show that $b^2 - 3ac = 0$.

7. When a model rocket is launched, it moves vertically upwards and its height (in metres) is given by the equation $h = 2t\sqrt{t} - \dfrac{1}{2}t^2$, where t is the time (in seconds) since take-off.

    (a) Find an expression for the velocity of the rocket.

    (b) Find the maximum velocity of the rocket.

    (c) Find the speed of the rocket when it returns to the ground.

8. Consider $f(x) = x^4 - x$.

    (a) Find the zeros of $f(x)$.

    (b) Find the region in which $y = f(x)$ is decreasing.

    (c) Solve the equation $f''(x) = 0$.

    (d) Find the region in which $y = f(x)$ is concave up.

    (e) Hence explain why $y = f(x)$ has no points of inflexion.

    (f) Sketch the curve $y = f(x)$.

Going for the top 7

1. The point (x, q) lies on the curve $y = x^2$.

    (a) Express q in terms of x.

    (b) Let d be the distance between (x, q) and the point $(0, 9)$.
        Write down an expression for d^2 in terms of x.

    (c) Find the value of x that gives the minimum value of d.

    (d) Hence write down the coordinates of the points on the curve $y = x^2$ that are closest to $(0, 9)$.

2. The point P lies on the curve $y = e^x$ with $x = p$, $p > 0$. O is the origin, and the line OP makes an angle of α with the positive x-axis. The tangent to the curve at P crosses the x-axis at Q

    (a) Let $p = 2$.

        (i) Find the value of α in degrees.

        (ii) Find the gradient of the tangent at P, and hence find the size of the angle $O\hat{P}Q$.

    (b) In this part of the question, p is not assigned any specific value.

        (i) Find, in terms of p, the equation of the tangent to the curve at P.

        (ii) Find the coordinates of Q.

    (c) A line with equation $y = kx$ is tangent to the curve. By considering your answer to part (b), find the value of k.

3. The point P lies on the curve $y = \dfrac{1}{x}$ with $x = p$, $p > 0$.

    (a) (i) Find, in terms of p, the gradient of the tangent to the curve at P.

        (ii) Show that the equation of the tangent is $p^2 y + x = 2p$.

    The tangent to the curve at P meets the y-axis at Q and meets the x-axis at R.

    (b) (i) Find the coordinates of Q and R.

        (ii) Calculate the area of the triangle OQR (where O is the origin).

    (c) Show that the distance QR is given by $2\sqrt{p^2 + p^{-2}}$.

    (d) Find the value of p that minimises the distance QR.

8 INTEGRATION

WHAT YOU NEED TO KNOW

- Integration is the reverse process of differentiation.

- The basic rules of integration:

  - For all rational $n \neq -1$, $\int x^n \, dx = \dfrac{x^{n+1}}{n+1} + C$, where C is the constant of integration.

  - $\int kf(x) \, dx = k \int f(x) \, dx$ for any constant k.

  - $\int f(x) + g(x) \, dx = \int f(x) \, dx + \int g(x) \, dx$

- Definite integration deals with integration between two points.

  - If $\int f(x) \, dx = F(x)$ then $\int_a^b f(x) \, dx = F(b) - F(a)$, where a and b are the limits of integration.

  - $\int_a^b f(x) \, dx + \int_b^c f(x) \, dx = \int_a^c f(x) \, dx$

  - $\int_a^b f(x) \, dx = -\int_b^a f(x) \, dx$

- Not all integrals are given in the Formula booklet. The table on the right lists some useful versions to learn.

Function	Integral
$\dfrac{1}{x}$	$\ln x + C$
$\sin x$	$-\cos x + C$
$\cos x$	$\sin x + C$
e^x	$e^x + C$

Function	Integral
$\dfrac{1}{ax+b}$	$\dfrac{1}{a}\ln(ax+b) + C$
$\sin kx$	$-\dfrac{1}{k}\cos kx + C$
$\cos kx$	$\dfrac{1}{k}\sin kx + C$
e^{kx}	$\dfrac{1}{k}e^{kx} + C$

- Integration by substitution can be useful when there is a composite function and (a multiple of) the derivative of the inner part of that function. (This can also be achieved by simply reversing the chain rule.)

  - $\int g'(x) f'(g(x)) \, dx = f(g(x)) + C$

- Integration can be used to calculate areas:

  - The area between a curve and the x-axis from $x = a$ to $x = b$ is given by $\int_a^b y\,dx$.

  - The area between two curves is given by $\int_a^b f(x) - g(x)\,dx$, where $f(x) > g(x)$ and a and b are the intersection points.

- The volume of revolution is given by $V = \int_a^b \pi y^2\,dx$ for rotation around the x-axis from $x = a$ to $x = b$.

- Integrate with respect to time to change an expression for acceleration (a) into an expression for velocity (v) and then into one for displacement (s):

  - $v = \int a\,dt$

  - $s = \int v\,dt$

    - The displacement between times t_1 and t_2 is $\int_{t_1}^{t_2} v\,dt$.

    - The distance travelled between times t_1 and t_2 is $\int_{t_1}^{t_2} |v|\,dt$.

⚠ EXAM TIPS AND COMMON ERRORS

- Don't forget the '+ C' for indefinite integration – it is part of the answer and you must write it every time. However, it can be ignored for definite integration.

- Make sure you know how to use your calculator to evaluate definite integrals. You can also check your answer on the calculator when you are asked to find the exact value of the integral.

- Always look out for integrals which are in the Formula booklet.

- You cannot integrate products or quotients by integrating each part separately.

- When integrating fractions, always check whether the numerator is the derivative of the denominator. If this is the case, the answer is the natural logarithm of the denominator:

$$\int \frac{f'(x)}{f(x)}\,dx = \ln f(x) + C.$$

- You may have to simplify a fraction before integrating. This can be achieved by splitting into separate fractions or by polynomial division (or equivalent method).

8.1 INTEGRATING EXPRESSIONS

WORKED EXAMPLE 8.1

Find $\int \dfrac{3-2\sqrt{x}}{5x}\,dx$.

$$\int \frac{3-2\sqrt{x}}{5x}\,dx = \int \frac{3}{5}\frac{1}{x} - \frac{2}{5}\frac{\sqrt{x}}{x}\,dx$$

$$= \int \frac{3}{5}\frac{1}{x} - \frac{2}{5}x^{-\frac{1}{2}}\,dx$$

> Use algebra to rewrite the expression in a form which can be integrated. Fractions can be split up and the rules of indices applied.

$$= \frac{3}{5}\int \frac{1}{x}\,dx - \frac{2}{5}\int x^{-\frac{1}{2}}\,dx$$

> We can split up the integral of a difference and move the constants outside the integrals.

$$= \frac{3}{5}\ln x - \frac{2}{5}\frac{x^{\frac{1}{2}}}{\frac{1}{2}} + c$$

 Even if there are two integrations, you only need to put in '+c' once.

$$= \frac{3}{5}\ln x - \frac{4}{5}x^{\frac{1}{2}} + c$$

Practice questions 8.1

1. Find $\int \dfrac{\sqrt{x}+1}{x}\,dx$.

2. Find $\int \sin 3x + e^{5x} + \sqrt{x}\,dx$.

3. Find $\int \left(3x + \sqrt{x}\right)^2 dx$.

 Remember to use the laws of algebra to prepare the integral first.

4. Find $\int (e^x + 3)^2\,dx$.

5. Find $\int (\sin x + \cos x)^2\,dx$.

 You will need to use a double angle formula in question 5. This is covered in Chapter 5.

6. Find $\int \tan 2x \cos 2x\,dx$.

7. Find $\int \left(x - \dfrac{1}{2x}\right)^2 dx$.

8.2 FINDING AREAS

WORKED EXAMPLE 8.2

Find the area enclosed between the curves $y = x^2 - 5x$ and $y = 7x - x^2$.

Using GDC:

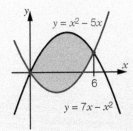

Sketch the graph on the GDC and use it to find the intersection points.

Intersections: $x = 0$ and $x = 6$

$$\therefore \text{Area} = \int_0^6 (7x - x^2) - (x^2 - 5x)\,dx$$

$$= \int_0^6 12x - 2x^2\,dx$$

$$= \left[6x^2 - \frac{2}{3}x^3 \right]_0^6 = 72$$

Write down the integral representing the area and carry out the definite integration.

The integral can be checked using your calculator.

Practice questions 8.2

8. Find the area enclosed by the curves $y = \sin x$ and $y = \cos x$ between $x = \dfrac{\pi}{4}$ and $x = \dfrac{3\pi}{4}$.

9. The diagram shows the graphs of $y = x^2 - 4x + 3$ and $y = 6x - 5 - x^2$. Find the shaded area.

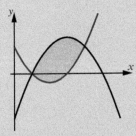

10. The diagram shows the graph of $y = x^2 - 3x + 2$. Find the shaded area.

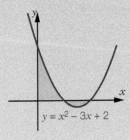

8.3 VOLUMES OF REVOLUTION

WORKED EXAMPLE 8.3

Find the exact volume of revolution when the curve $y = x^2$ between $x = 1$ and $x = 3$ is rotated $360°$ about the x-axis.

$$\text{Volume} = \pi \int_1^3 y^2 \, dx$$

$$= \pi \int_1^3 \left(x^2\right)^2 dx$$

$$= \pi \int_1^3 x^4 \, dx$$

$$= \pi \left[\frac{x^5}{5}\right]_1^3$$

$$= \frac{242\pi}{5}$$

To visualise this, sketch a graph.

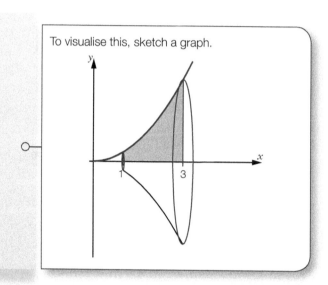

Practice questions 8.3

11. Find the exact volume generated when $y = e^{3x}$, for $1 < x < 3$, is rotated $360°$ around the x-axis.

12. The diagram shows the graph of $y = x^2$. Find the volume generated when the shaded area is rotated 2π radians about the y-axis.

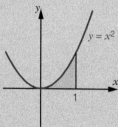

13. The region bounded by the curve $y = x^2$, the line $y = 6 - x$ and the x-axis is shown below.

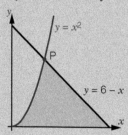

    (a) Find the coordinates of the point P.
    (b) Hence find the volume generated when this region is rotated $360°$ around the x-axis.

8.4 KINEMATICS

WORKED EXAMPLE 8.4

The velocity, $v\,\mathrm{m\,s^{-1}}$, of a ball is given by $v = t^2 - 4t$, where t is measured in seconds. Initially the displacement is zero.

(a) Find the displacement when $t = 10$.

(b) Find the distance travelled in the first 10 seconds.

(a) $s = \displaystyle\int_0^{10} v\,dt$

      $= \displaystyle\int_0^{10} t^2 - 4t\,dt$

      $= \dfrac{400}{3}\,m$ (from GDC)

> 'Initially' means when $t = 0$, so we integrate between the limits $t = 0$ and $t = 10$.

(b) Distance $= \displaystyle\int_0^{10} |v|\,dt$

      $= \displaystyle\int_0^{10} |t^2 - 4t|\,dt$

      $= \dfrac{464}{3}\,m$ (from GDC)

> To find the distance travelled, we need to take the modulus before integrating, as the velocity may be negative for part of the journey.

> ⚠ To answer a question of this type without a GDC, first sketch the velocity–time graph and then separate out the parts above the axis and below the axis.

Practice questions 8.4

14. The acceleration of a car for $0 \le t \le 5$ is modelled by $a = 5\left(1 - e^{-2t}\right)$, where a is measured in $\mathrm{m\,s^{-2}}$ and t is measured in seconds. The car is initially at rest.
  (a) Find the velocity of the car after t seconds.
  (b) Find the displacement from the initial position after t seconds.
  (c) Find the maximum velocity of the car.

15. The velocity of a wave is modelled by $v = \cos 5t$. When $t = 0$, $s = 0$. Show that the acceleration a and displacement s are related by $a = ks$, where k is a constant to be determined.

16. A bird has acceleration modelled by $2e^{-t}\,\mathrm{m\,s^{-2}}$ due north, where t is the time in seconds. The bird is initially travelling with speed $8\,\mathrm{m\,s^{-1}}$ north and is $100\,\mathrm{m}$ south of a tree.
  (a) Find an expression for the velocity of the bird at time t.
  (b) According to the model, when does the bird reach the tree?

8.5 INTEGRATION BY SUBSTITUTION

WORKED EXAMPLE 8.5

Find $\int x^2 \cos(x^3 + 2)\,dx$.

Let $u = x^3 + 2$. Then $\dfrac{du}{dx} = 3x^2$

$\therefore dx = \dfrac{du}{3x^2}$

$\int x^2 \cos(x^3 + 2)\,dx = \int x^2 \cos u\, \dfrac{du}{3x^2}$

$\qquad\qquad\qquad = \dfrac{1}{3}\int \cos u\, du$

$\qquad\qquad\qquad = \dfrac{1}{3}\sin u + c$

$\qquad\qquad\qquad = \dfrac{1}{3}\sin(x^3 + 2) + c$

 The integral involves a composite function, so integration by substitution will be useful. Choose $u = $ 'inner function'.

 Don't forget to substitute for dx too.

Write the answer in terms of x.

Questions of this type can be done more directly by recognising that the x^2 factor is a multiple of the derivative of $x^3 + 2$ and then simply reversing the chain rule. This is an acceptable alternative method which would also gain full marks.

Practice questions 8.5

17. Find $\int x^3 \, e^{x^4}\,dx$.

18. Find $\int \cos x\, e^{\sin x}\,dx$.

19. Find $\int 2x\sqrt{x^2 - 4}\,dx$.

20. By using the substitution $u = \cos x$, find $\int \tan x\,dx$.

21. Find $\int \dfrac{\cos x}{\sin x + 4}\,dx$.

22. Find $\int \dfrac{\ln x}{x}\,dx$.

Mixed practice 8

1. Find: (a) $\int \sqrt{e^x}\, dx$ (b) $\int_0^{\pi/10} 3\cos(5x)\, dx$

2. Use the substitution $u = e^x + 1$ to find the exact value of $\int_0^{\ln 2} \dfrac{e^x}{\sqrt{e^x + 1}}\, dx$.

3. Find the area enclosed by the curves $y = 4\sqrt{x}$ and $y = \dfrac{1}{2}x^2$.

4. Evaluate $\int_0^1 e^{\sin x}\, dx$.

5. A ball's velocity, $v\,\mathrm{m\,s^{-1}}$, after time t seconds is given by $v = 3\sin t$.

   (a) Find the displacement of the ball from the initial position when $t = \dfrac{3\pi}{2}$ s.

   (b) After how long has the ball reached its maximum displacement in the first $\dfrac{3\pi}{2}$ seconds?

   (c) Find the distance travelled by the ball in the first $\dfrac{3\pi}{2}$ seconds.

6. Find a if $\int_0^a \sin 2x\, dx = \dfrac{3}{4},\ 0 < a \le \pi$.

7. A curve has gradient given by the equation $\dfrac{dy}{dx} = \dfrac{2}{3x+1}$, and when $x = 0$, $y = 0$.

   Find the value of x for which $y = 2$.

8. An object moves with acceleration $a = \dfrac{1}{(t-2)^2}$, where t is time and $t \ge 3$. When $t = 3$, the velocity of the object is 5 and its displacement from the origin is 20.

   (a) Find an expression for the velocity of the object.

   (b) Find the displacement of the object from the origin when $t = 5$.

9. (a) Using an appropriate double angle formula, show that $\sin^2 \theta = \dfrac{1}{2}(1 - \cos 2\theta)$.

   (b) Hence find $\int \sin^2 \theta\, d\theta$.

   The region R is enclosed by the x-axis and the graph of $y = \sin x$ between $x = 0$ and $x = \pi$.

   (c) Find the area of R.

   (d) Find the exact volume of the solid generated when R is rotated $360°$ around the x-axis.

   (e) Use the substitution $u = \cos x$ to find $\int \sin^3 x\, dx$.

Going for the top 8

1. (a) Find the value of C such that $x - 2 + \dfrac{C}{x+2} = \dfrac{x^2+3}{x+2}$.

   (b) Hence find $\displaystyle\int \dfrac{x^2+3}{x+2}\,\mathrm{d}x$.

2. Using a suitable substitution, show that $\displaystyle\int_{e^3}^{e^{12}} \dfrac{1}{2x\ln x}\,\mathrm{d}x = \ln 2$.

3. The region between the curves $y = x + \dfrac{2}{x}$ and $y = 5 - x$ is labelled R.

   (a) Use integration to find the area of R.

   (b) Find the volume generated when R is rotated a full turn around the x-axis.

4. Use a suitable substitution to find $\displaystyle\int \sin^4\theta\cos\theta\,\mathrm{d}\theta$.

5. (a) Use a double angle formula to show that $\displaystyle\int \cos^2 x - \sin^2 x\,\mathrm{d}x = \dfrac{1}{2}\sin 2x + c$.

   The diagram shows the graphs of $y = \cos x$ and $y = \sin x$.

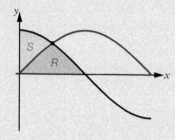

   (b) Find the x-coordinate of the point of intersection shown in the diagram.

   (c) Find the exact area of the region labelled R.

   (d) Find the volume of the solid generated when the region labelled S is rotated around the x-axis.

9 DESCRIPTIVE STATISTICS

- Grouping a large sample of data into groups (or classes) makes it easier to summarise.
  - The upper and lower class boundaries are the largest and smallest data values that would be included in that group.
  - The mid-interval value is the mean of the upper and lower class boundaries.
  - If the data are discrete, the upper and lower class boundaries are as shown in the grouped frequency table. For example, a group 12–15 includes the values 12, 13, 14 and 15, and the mid-interval value is 13.5.
  - If the data are continuous but have been rounded, the class boundaries need to be adjusted. For example, if lengths have been rounded to the nearest metre, then 12–15 means that $11.5 \leq \text{length} < 15.5$, and the mid-interval value is 13.5.
  - Age is rounded differently from other measurements. For example, 12–15 years means $12 \leq \text{age} < 16$, and the mid-interval value is 14.
- Three measures of central tendency are the mean, median and mode.
  - The mean is the sum of all the data values divided by the total number of data items:

$$\bar{x} = \frac{\sum_{i=1}^{n} f_i x_i}{\sum_{i=1}^{n} f_i}$$

  To find the mean for grouped data, assume that every item is at the centre of its group; that is, use the mid-interval value of the group for x_i.
  - The median is the middle of a data set whose values have been arranged in order of size.
  - The mode is the most common data value.
- There are three ways of measuring how spread out a data set is: range, interquartile range and standard deviation.
  - The range is the difference between the largest value and the smallest value.
  - The interquartile range (IQR) is the difference between the upper quartile (Q_3) and the lower quartile (Q_1): $\text{IQR} = Q_3 - Q_1$.
    - To find the quartiles, first split the data set (with numbers arranged in order) into two halves. The lower quartile is the middle value of the bottom half. The upper quartile is the middle value of the top half.
  - Standard deviation can be calculated on a GDC.
    - To estimate the standard deviation or mean of grouped data, assume that every item is at the centre of its group; that is, use the mid-interval values for calculations.

- The measures of central tendency and spread can be affected when constant changes are applied to the original data, for example, to convert units or to simplify calculations.

  - If every data item is increased (or decreased) by the value x, all measures of the centre of the data will also increase (or decrease) by x, while all measures of the spread of the data will remain unchanged.

  - If every data item is multiplied (or divided) by the value x, all measures of the centre of the data and all measures of the spread of the data will also be multiplied (or divided) by x.

    - If the standard deviation is multiplied by x, the variance is multiplied by x^2.

- In a histogram, there should be no gap between the bars, and the height of each bar is equal to the frequency of that group.

- Cumulative frequency is the total frequency up to a certain data value.

  - To draw a cumulative frequency curve:

    - Start the curve from a point on the x-axis corresponding to the lower boundary of the first group.

    - For each group, plot the cumulative frequency against the upper boundary of the group. Join up the points with a smooth increasing curve.

  - The cumulative frequency curve can be used to find the median and quartiles.

    - For the median, the value on the y-axis will be at $\dfrac{\text{total frequency}}{2}$; for the lower quartile, the value will be at $\dfrac{\text{total frequency}}{4}$; and for the upper quartile it will be at $\dfrac{3 \times \text{total frequency}}{4}$.

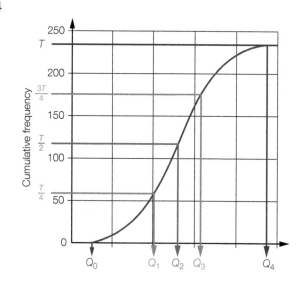

- A box and whisker diagram summarises five important values from a set of data: the smallest value (Q_0), the lower quartile (Q_1), the median (Q_2), the upper quartile (Q_3) and the largest value (Q_4).

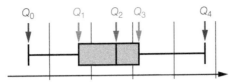

- Each 'whisker' and each half of the 'box' contains a quarter of all the data.
- To compare two sets of data:
  - Use the mean and median to compare 'averages'.
  - Use the standard deviation and IQR to compare the spread of the data.
    - The data set with the smaller spread can be described as 'less varied' or 'more consistent'.
- Correlation is a relationship between two variables. Correlation does not imply causation.
  - Pearson's product–moment correlation coefficient, r, is a measure of the linear correlation between two variables. It can take any value between -1 and $+1$.
    - A value of r close to -1 shows that there is a strong negative correlation; a value close to $+1$ shows that there is a strong positive correlation; and a value close to 0 shows no linear correlation.
- A regression line describes a linear relationship between two variables.
  - The variable plotted on the x-axis is the independent, or controlled, variable. The variable plotted on the y-axis is the dependent variable.
  - The regression line always passes through the mean point of the data.
  - The regression line can be used to estimate values that are not in the original data. Such an estimate is reliable only if the correlation is strong and the values of the variables are within the range of the original data.
- The correlation coefficient and the equation of the regression line can be found using a GDC.

 EXAM TIPS AND COMMON ERRORS

- When using a calculator to find statistics, make sure you show the numbers that you are using.
- A common error when drawing a cumulative frequency curve is forgetting to plot the first point, which corresponds to zero frequency at the lower boundary of the first group.
- In a box and whisker diagram, the width of the box represents the interquartile range, **not** the number of data values in that range; it is a common mistake to say that a wider box 'contains more people'. When comparing two box and whisker diagrams, always make it clear whether it is the width of the box or its position that is being referred to. 'Higher IQR' should mean that the box is wider, not that the quartiles are larger.
- If a question asks you to find the equation of a 'suitable regression line', you should first identify the independent variable and enter the corresponding values into the X list on your calculator. The data given in the question may be presented the wrong way round!

9.1 MEDIAN AND QUARTILES

WORKED EXAMPLE 9.1

Consider the following set of data:

14, 12, 18, 20, 17, 18, 12, 16, 15, x

(a) Given that the median of the data set is 16, find the value of x.

(b) Find the interquartile range of the data.

(a) 12, 12, 14, 15, 16, 17, 18, 18, 20

Put the nine known numbers in order of size. Including x, there are 10 numbers, so the median is the mean of the 5th and 6th values.

The median is the mean of 16 and x.
Since the median is 16, $x = 16$.

The known number 16 is either the 5th or the 6th value, so the median is the mean of 16 and another number. Since we are given that the median is 16, the other number must also be 16.

(b) Lower half: 12, 12, **14**, 15, 16 $\Rightarrow Q_1 = 14$
Upper half: 16, 17, **18**, 18, 20 $\Rightarrow Q_3 = 18$
IQR $= 18 - 14 = 4$

To find the quartiles Q_1 and Q_3, divide the data set (with numbers in order) into two halves and find the median for each half. Then use the formula IQR $= Q_3 - Q_1$.

Practice questions 9.1

1. The following data set contains 13 numbers:
2, 4, 5, 10, 12, 14, 14, 16, 18, 21, 23, 23, 25
(a) Explain why the lower quartile is 7.5.
(b) Find the interquartile range of the data.

2. The median of the numbers $x - 2$, $x + 1$, $x + 3$, $x + 6$, $x + 7$, $x + 10$ is 6.5.
Find the value of x.

3. Find the median and the interquartile range of the grades summarised in the table:

Grade	2	3	4	5	6	7
Frequency	5	12	36	42	27	19

When entering frequency table data into your GDC, check that the value of n is what you expect; in this case 141.

9.2 MEAN, STANDARD DEVIATION AND FREQUENCY HISTOGRAMS

WORKED EXAMPLE 9.2

The histogram below shows the masses of some dogs (in kilograms).

(a) How many dogs were included in the sample?

(b) What is the modal group?

(c) Estimate the standard deviation of the masses.

(d) All the masses are converted into pounds (lb), where 1 kg = 2.205 lb. Find the standard deviation of the masses measured in pounds.

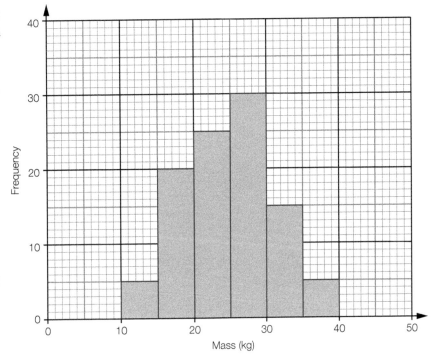

Mass (kg)

(a) Total frequency = 5 + 20 + 25 + 30 + 15 + 5
= 100

There were 100 dogs in the sample.

> The number of dogs in each group is represented by the height of the histogram bar.

(b) The modal group is 25–30 kg.

> The modal group is the data class with the highest bar.

(c)

Mid-interval value	12.5	17.5	22.5	27.5	32.5	37.5
Frequency	5	20	25	30	15	5

(From GDC) Standard deviation = 6.22 (3 SF)

> To estimate the standard deviation, we need to use the mid-interval value of each group, which is the mean of the lower and upper class boundaries of that group. For the first group, the mid-interval value is $\dfrac{10+15}{2} = 12.5$.

> A GDC gives two standard deviation values. You need to use the smaller one.

(d) Standard deviation in pounds
= 6.22 × 2.205 = 13.7 lb

> When all the data values are multiplied by a constant, the standard deviation is multiplied by the same constant.

Practice questions 9.2

4. The mean of the numbers 4, 5, 3, 3, 5 and k is 3.5.
   (a) Find the value of k.
   (b) Find the standard deviation of the numbers.

5. The frequency table shows the heights of 26 trees. The mean height is 6.5 m.

Height (m)	3	5	y	10
Frequency	4	x	11	5

   (a) Find the values of x and y.
   (b) Calculate the standard deviation of the heights.

6. The heights of 50 buildings, rounded to the nearest metre, are summarised in the following table:

Height (m)	12–17	18–23	24–29	30–35
Frequency	12	14	16	8

   (a) Write down the upper and lower boundaries of the 24–29 class.
   (b) Draw a histogram to represent the data.
   (c) Find the mean and standard deviation of the heights.
   (d) The heights of 50 hills are recorded in the following table:

Height (m)	512–517	518–523	524–529	530–535
Frequency	12	14	16	8

   Write down the mean and standard deviation of these heights.

7. The ages of 35 teachers are recorded in the table:

Age (years)	21–30	31–40	41–50	51–60	61–70
Frequency	5	9	11	7	3

   (a) Write down the upper and lower boundaries of the 21–30 group and find its mid-interval value.
   (b) Draw a histogram to represent the data.
   (c) Find the mean and standard deviation of the ages.

9.3 CUMULATIVE FREQUENCY

WORKED EXAMPLE 9.3

The cumulative frequency curve below represents the heights of 36 children.

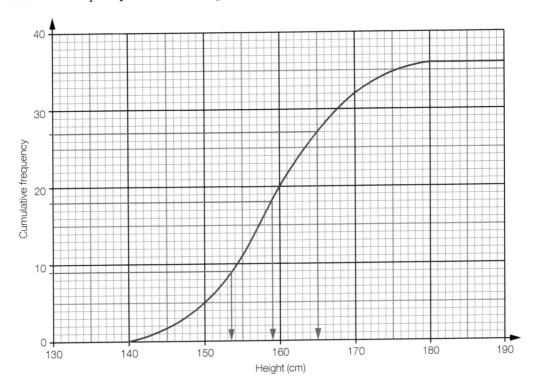

Estimate the median and interquartile range of the heights.

Median ≈ 159 cm

$Q_1 \approx 153$ cm

$Q_3 \approx 165$ cm

IQR ≈ 165 − 153 = 12 cm

$\frac{1}{2} \times 36 = 18; \frac{1}{4} \times 36 = 9; \frac{3}{4} \times 36 = 27$

To estimate the median, draw a line horizontally across from 18 on the vertical axis to the curve and then vertically down until it meets the horizontal axis. Similarly find Q_1 and Q_3, and hence calculate the IQR.

Practice questions 9.3

8. The ages of students at a school are summarised in the table.

(a) Draw a cumulative frequency curve to represent the information.

(b) Find the median age.

(c) The oldest 10% of the students are older than m years. Find the value of m.

Age (years)	Frequency
4–7	20
8–11	40
12–15	80
16–19	60

9. Use the cumulative frequency curve below to complete the frequency table.

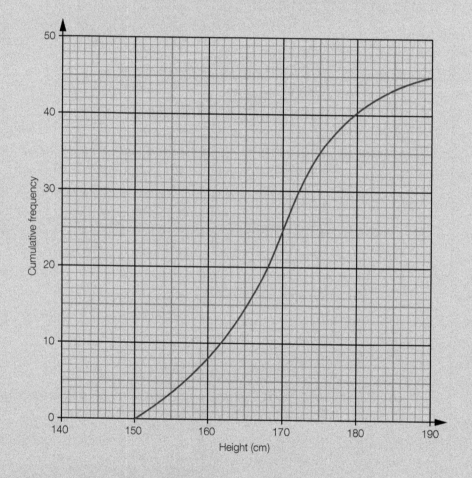

Height, h (cm)	Frequency
$150 \leq h < 160$	8
$160 \leq h < 168$	
$168 \leq h < 175$	
$175 \leq h < 190$	

WORKED EXAMPLE 9.4

The English test marks (in %) of a group of students are summarised in the following diagram.

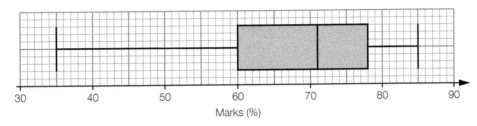

Marks (%)

The marks for the same group of students on a Mathematics test had the following features:

lowest mark 24%, highest mark 93%, lower quartile 43%, median 62%, upper quartile 72%.

(a) Draw a box and whisker diagram for the Mathematics marks.

(b) Compare the performance of the group in the two tests.

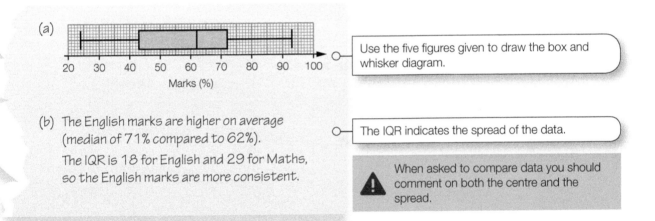

(a)

Marks (%)

Use the five figures given to draw the box and whisker diagram.

(b) The English marks are higher on average (median of 71% compared to 62%).

The IQR is 18 for English and 29 for Maths, so the English marks are more consistent.

The IQR indicates the spread of the data.

⚠ When asked to compare data you should comment on both the centre and the spread.

Practice questions 9.4

10. Draw a box and whisker diagram to represent the following data:

Smallest	Largest	Mean	Standard deviation	Median	Q_1	Q_3
12	45	31.6	5.5	35	23	42

11. Two schools took part in a cross-country race. The times are summarised in the diagram below. Write three comments comparing the times of the students from the two schools.

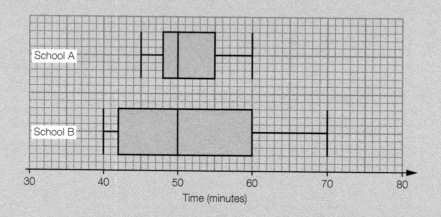

When comparing two sets of data, always refer to the context of the question. For example, instead of 'the men have a higher mean' you should say 'the men are older on average'.

9.5 CORRELATION AND REGRESSION

WORKED EXAMPLE 9.5

The following table shows the outside temperature and the amount of money taken by a hot chocolate machine, recorded on nine randomly selected days.

Temperature, T (°C)	23	12	5	8	16	−4	1	−2	9
Sales, S ($)	72	112	142	148	85	161	170	132	85

(a) Calculate the value of Pearson's product–moment correlation coefficient for the data, and comment on the value of r.

(b) A regression line is to be used to describe the relationship between the temperature and sales of hot chocolate. Identify the independent variable and find the equation of a suitable regression line.

(c) Estimate the outside temperature on a day when the total sales of hot chocolate are $120.

(d) The regression line is used to estimate the sales of hot chocolate on a day when the outside temperature is 35°C. Comment on the reliability of this estimate.

(a) From GDC: $r = -0.832$

The value of r is close to -1, indicating that there is strong negative correlation between the outside temperature and sales of hot chocolate. As the outside temperature decreases, the sales of hot chocolate increase.

⚠ You must refer to the context of the question when interpreting the value of r.

(b) The independent variable is the temperature, T.
From GDC: $S = -3.44T + 149$

The temperature may affect the sales, but the sales cannot affect the temperature. Hence, the regression line should be of S on T. Put the T values into the X list on the GDC.

(c) When $S = 120$: $120 = -3.44T + 149$

$3.44T = 29$

$\therefore T = 8.43$°C

Substitute $S = 120$ into the equation of the regression line from part (b).

(d) The estimate would be unreliable, because 35°C is outside the range of the data values already collected.

⚠ Do not use the regression line to predict new values if the predicted values are outside the range of the data values already collected (this is called extrapolation).

Practice questions 9.5

12. The following table shows the English and History test marks obtained by a group of eight students.

History mark (%)	62	48	82	71	53	67	90	56
English mark (%)	54	71	52	46	85	72	76	82

(a) Calculate Pearson's product–moment correlation coefficient, r, for the data, and comment on what the value of r suggests about the relationship between the History and English marks.

(b) Another student scored 63% in the English test but missed the History test. State, with a reason, whether a regression line could be used to estimate the History mark she would have got.

13. Ten athletes recorded their average number of hours of training per week (h) and their 100 m time in a competition (t seconds):

h (hours)	12	16	8	21	10	14	6	17	21	9
t (seconds)	12.7	11.3	12.9	10.7	11.2	12.2	13.1	11.4	11.1	12.6

(a) Find the product–moment correlation coefficient between the number of training hours and the 100 m time.

(b) Identify the independent variable and find the equation of a suitable regression line.

(c) Mario achieved a time of 11.6 seconds in the race.
  (i) Use your regression line to estimate how many hours of training per week he did.
  (ii) Comment on the reliability of your estimate.

(d) Jack says that the calculations above prove that he will decrease his 100 m time if he trains more hours per week. State, with a reason, whether he is right.

14. The managers of two companies want to see whether there is any correlation between the age (A) and salary (S) of their employees. The employees in the sample are aged between 16 and 42.

(a) The first manager finds that the correlation coefficient of the data A and S is 0.872 and the equation of the regression line is $S = 1000A + 2000$. He uses the equation to estimate salaries for given ages. For each of the values of A below, either estimate the salary or explain why the estimate would not be reliable.
  (i) $A = 37$ (ii) $A = 56$

(b) The second manager finds that the correlation coefficient is -0.217 and the equation of the regression line is $S = -502A + 2720$. He uses this equation to estimate the salary of a 25-year-old employee. Comment on the reliability of this estimate, giving reasons for your answer.

Mixed practice 9

1. The results of a Physics exam for two different schools are summarised in the table below:

	Lowest mark	Highest mark	Median	Lower quartile	Upper quartile
School 1	20	52	32	26	41
School 2	31	60	39	34	48

(a) Calculate the interquartile ranges of the marks for the two schools.

(b) Draw two box and whisker diagrams to represent the results.

(c) Describe one similarity and one difference between the two schools' results.

2. The table shows the History grades of IB students at a college:

Grade	3	4	5	6	7
Number of students	4	12	x	17	9

(a) Given that the mean grade is 5.23 (to three significant figures), find the value of x.

(b) Find the median grade.

3. Match each diagram with the most appropriate value of Pearson's product–moment correlation coefficient.

(a) $r = 0.932$ (b) $r = -0.561$ (c) $r = 0.125$

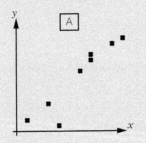 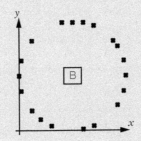 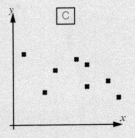

4. The following table shows the masses (m kg) and lengths (l cm) of 11 babies.

m	2.7	4.3	3.4	2.9	3.6	4.7	4.1	3.3	3.1	4.3	3.7
l	48	55	52	47	51	56	53	51	50	51	49

(a) Write down the correlation coefficient, r, of the data.

(b) Find the equation of the regression line of l on m.

(c) Use your equation to estimate the length of a baby whose mass is 3.2 kg.

(d) Can the regression line be reliably used to estimate the length of a baby of mass 5.6 kg? Explain your answer.

5. The ages of employees at a company are summarised in the cumulative frequency table. The youngest employee is 16 years old.

Age (years)	Cumulative frequency
≤ 26	12
≤ 36	46
≤ 46	82
≤ 56	90

(a) Draw a histogram to represent the data.

(b) Estimate the mean and standard deviation of the ages.

6. All athletes in a club competed in a long jump competition. Their results are shown in the histogram:

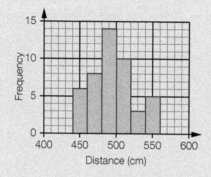

(a) Use the histogram to complete the frequency table, where the distances have been rounded to the nearest centimetre.

Distance (cm)	441–460	461–480	481–500	501–520	521–540	541–560
Frequency						

(b) Estimate the mean and standard deviation of the distances.

(c) The distances are converted from centimetres to inches (1 inch = 2.54 cm). Find the mean and standard deviation of the distances measured in inches.

(d) Draw a cumulative frequency curve for the original data.

(e) (i) Find the percentage of athletes who jumped further than 4.80 m.

(ii) Two athletes are selected at random. What is the probability that they both jumped further than 4.80 m?

(f) The top 20% of athletes will qualify for a regional competition. Estimate the minimum distance required for qualification.

Going for the top 9

1. The frequency table summarises 36 pieces of data with mean $\dfrac{47}{9}$. Find the values of x and y.

Value	4	5	6	7
Frequency	9	13	x	y

2. The set of numbers 3, 2, 3, 7, 10, 5, 7, 12, x, y, z has mode 3, median 6 and mean 6. Find the values of x, y and z.

3. The histogram shows the times a group of 55 students took to complete their homework.

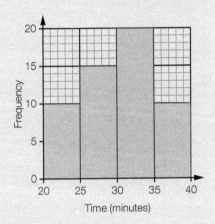

(a) Estimate the number of students who took:

   (i)   more than 25 minutes

   (ii) more than 37 minutes.

(b) Given that a student took more than 25 minutes to complete their homework, find the probability that they took more than 37 minutes.

4. A sports club recorded the amount of money they spent on advertising (a dollars) over a period of several years together with the number of members (M). They found that there is a strong correlation between the two, and that the equation of the regression line is $M = 123 + 2.6a$.

(a) How much money does the club need to spend on advertising if they want 250 members?

(b) What does the number 123 in the equation of the regression line represent?

5. The test scores of a group of 120 students are shown in the frequency histogram below.

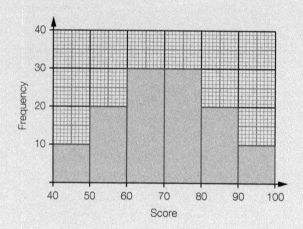

(a) Estimate the mean and the standard deviation of the test scores.

(b) The scores are converted to the standard score scale by using the following formula:

$$\text{standardised score} = \frac{\text{original score} - 30}{10}$$

Find the mean and standard deviation of the standardised scores.

(c) Estimate the number of students whose test score is within one standard deviation of the mean.

(d) **Hence** explain whether a normal distribution would be a suitable model for these test scores.

 Normal distribution is covered in Chapter 10.

10 PROBABILITY

WHAT YOU NEED TO KNOW

- The probability of an event can be found by listing or counting all possible outcomes.

- The probabilities of combined events are often best found using Venn diagrams or tree diagrams.

  - Venn diagrams are often useful when the question involves the union of events.

    - The union of events A and B can be found using the formula:

      $$P(A \cup B) = P(A) + P(B) - P(A \cap B)$$

    - Mutually exclusive events cannot happen together: $P(A \cap B) = 0$. For mutually exclusive events, the above formula becomes:

      $$P(A \cup B) = P(A) + P(B).$$

  - Tree diagrams can be used when the question involves a sequence of events.

- The probability of event B happening given that an event A has already happened is known as conditional probability and is given by:

  $$P(B \mid A) = \frac{P(A \cap B)}{P(A)}$$

  - Conditional probabilities are usually best represented on a tree diagram. A conditional probability can also be found from a Venn diagram, by considering only the part of the diagram corresponding to the given condition.

  - Events A and B are independent if $P(A \mid B) = P(A)$ or, equivalently, $P(A \cap B) = P(A)\,P(B)$.

- A discrete random variable can be described by its probability distribution, which is the list of all possible values and their probabilities.

  - The total of all the probabilities must always equal 1.

  - The expected value is $E(X) = \mu = \sum_{x} x\, P(X = x).$

- If there is a fixed number of trials (n) with constant and independent probability of success (p) in each trial, the number of successes follows a binomial distribution: $X \sim B(n, p)$.

  - $E(X) = np$

  - $\text{Var}(X) = np(1 - p)$

  - A calculator can be used to find probabilities of the form $P(X = k)$ and $P(X \le k)$. Other probabilities can be found from these; for example:

    - $P(X \ge 7) = 1 - P(X \le 6)$

    - $P(5 \le X < 9) = P(X \le 8) - P(X \le 4)$

- The normal distribution is determined by its mean (μ) and variance (σ^2): $X \sim N(\mu, \sigma^2)$.
  - Normal probabilities are found using a calculator.
  - On a normal distribution diagram, the shaded area represents the probability.
  - The inverse normal distribution is used to find the value of x that corresponds to a given cumulative probability: $p = P(X \leq x)$.
  - If μ or σ is unknown, use the standard normal distribution to replace the values of X with their Z-scores, $z = \dfrac{x - \mu}{\sigma}$, which satisfy $Z \sim N(0, 1)$.

⚠ EXAM TIPS AND COMMON ERRORS

- Make sure that you do not confuse standard deviation and variance, especially when working with the normal distribution.

- When interpreting probability questions, pay particular attention to whether the required probability is conditional or not.

- Most conditional probability questions can be answered by considering a tree diagram or Venn diagram. Similarly, most questions involving $P(A \cup B)$ can be answered from a Venn diagram, without having to use the formula.

- With tree diagrams, remember to multiply along the branches and add down the branches.

- When working with the binomial distribution, make sure that you do not confuse single probabilities with cumulative probabilities (pdf and cdf on your calculator). When finding a cumulative probability, always check whether the endpoints should be included.

- In normal distribution questions, always start by drawing a diagram and shading the required probability.

- When you use your GDC to find binomial and normal probabilities, you must write the results using the correct mathematical notation, **not** calculator notation.

10.1 CALCULATING PROBABILITIES BY CONSIDERING POSSIBLE OUTCOMES

WORKED EXAMPLE 10.1

A cubical die has the numbers 1, 1, 2, 3, 3, 3 written on its faces. The die is rolled twice. Find the probability that the sum of the two scores is greater than 4.

The possible sums are $S = 5$ and $S = 6$.

$$P(S=6)=P(3 \cap 3)$$
$$= P(3) \times P(3)$$
$$= \frac{1}{2} \times \frac{1}{2} = \frac{1}{4}$$

> The scores on the two rolls of the die are independent, so $P(3 \cap 3) = P(3) \times P(3)$.

There are two ways of getting a sum of 5: $2 + 3$ and $3 + 2$.

$$P(S=5)=P\big((2 \text{ then } 3) \text{ OR } (3 \text{ then } 2)\big)$$
$$= \big(P(2) \times P(3)\big) + \big(P(3) \times P(2)\big)$$
$$= \frac{1}{6} \times \frac{1}{2} + \frac{1}{2} \times \frac{1}{6} = \frac{1}{6}$$

> The event 'a 2 followed by a 3' has to be counted as a separate event from 'a 3 followed by a 2'.

The total probability is $\dfrac{1}{4} + \dfrac{1}{6} = \dfrac{5}{12}$.

> The two events $S = 5$ and $S = 6$ are mutually exclusive, so we use $P(A \cup B) = P(A) + P(B)$.

Practice questions 10.1

1. The random variable X has probability distribution as shown in the table below.

x	1	2	3
$P(X = x)$	0.4	0.3	0.3

Find the probability that the sum of two independent observations of X is 4.

2. A standard cubical die is rolled twice.
   (a) Find the probability that the larger number is 5 (or that both are equal to 5).
   (b) Find the probability that the product of the scores is a multiple of 4.

3. A bag contains seven caramels and one chocolate. Three children, Peng, Quinn and Raul, take turns to pick a sweet out of the bag at random. If the sweet is a chocolate, they take it; if it is a caramel, they put it back and pass the bag around.
   (a) Find the probability that Raul gets the chocolate on his second turn.
   (b) Find the probability that the chocolate is still in the bag when it gets to Quinn for the fifth time.

10.2 VENN DIAGRAMS AND SET NOTATION

WORKED EXAMPLE 10.2

The events A and B are such that $P(A \cup B) = \dfrac{3}{5}$, $P(B) = \dfrac{2}{5}$ and $P(B \mid A) = \dfrac{3}{7}$.

(a) State, with a reason, whether A and B are independent.

(b) Find $P(A)$.

(a) The events A and B are not independent because $P(B \mid A) \neq P(B)$.

> For independent events, knowing that A has occurred has no impact on the probability of B occurring, i.e. $P(B \mid A) = P(B)$.

(b) Let $P(A \cap B) = x$. Then $P(B \cap A') = \dfrac{2}{5} - x$ and

$$P(A \cap B') = \dfrac{3}{5} - \left(\dfrac{2}{5} - x\right) - x = \dfrac{1}{5}$$

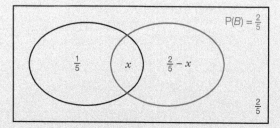

$P(B) = \frac{2}{5}$

$\frac{1}{5}$ x $\frac{2}{5} - x$

$\frac{2}{5}$

> ⚠ Venn diagrams are a useful way of representing information when you are given information about the union. Label the intersection x and work outwards.

$$P(B \mid A) = \dfrac{P(B \cap A)}{P(A)}$$

$$\therefore \dfrac{3}{7} = \dfrac{x}{\dfrac{1}{5} + x}$$

$$\Rightarrow x = \dfrac{3}{20}$$

Hence $P(A) = \dfrac{1}{5} + \dfrac{3}{20} = \dfrac{7}{20}$.

> For conditional probability we can use the formula $P(B \mid A) = \dfrac{P(B \cap A)}{P(A)}$.

Practice questions 10.2

4. Given that $P(B) = 0.5$, $P(A' \cap B') = 0.2$ and $P(B \mid A) = 0.4$, find $P(A)$.

5. All of the 100 students at a college take part in at least one of three activities: chess, basketball and singing. 10 play both chess and basketball, 12 play chess and sing, and 7 take part in both singing and basketball. 40 students play basketball, 62 play chess and 22 sing.

   (a) How many students take part in all three activities?

   (b) A student is chosen at random. Given that this student plays basketball and sings, what is the probability that she also plays chess?

10.3 TREE DIAGRAMS

WORKED EXAMPLE 10.3

The events A and B are such that $P(A) = \dfrac{1}{3}$, $P(B \mid A) = \dfrac{1}{3}$ and $P(B \mid A') = \dfrac{1}{2}$. Find $P(A \mid B)$.

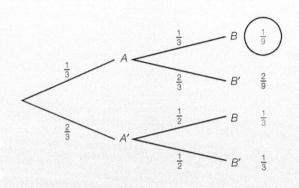

The conditional probability $B \mid A$ can be represented on a tree diagram as a branch after event A, and the conditional probability $B \mid A'$ can be represented as a branch after A'.

The remaining probabilities in the diagram can then be filled in.

$$P(A \mid B) = \frac{P(A \cap B)}{P(B)}$$

$$= \frac{\dfrac{1}{9}}{\dfrac{1}{9} + \dfrac{1}{3}} = \frac{1}{4}$$

We can find $P(A \mid B)$ using the formula

$$P(A \mid B) = \frac{P(A \cap B)}{P(B)}.$$

The probabilities shown in red correspond to event B happening, and the circled one is the probability of both A and B happening.

Practice questions 10.3

6. The events A and B are such that $P(B \mid A) = 0.2$, $P(B' \mid A') = 0.3$, $P(A \mid B) = 0.4$ and $P(A) = x$.

(a) Complete the following tree diagram.

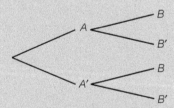

(b) Find the value of x.

(c) State, with a reason, whether the events A and B are independent.

7. Given that $P(B) = 0.3$, $P(A \mid B) = 0.6$ and $P(A \mid B') = 0.4$, find $P(B \mid A)$.

8. Every morning I either walk or cycle to school, with equal probability. If I walk, the probability that I am late is 0.2. If I cycle, the probability that I am late is 0.4. Given that I was late for school yesterday, what is the probability that I walked?

9. A box contains 17 yellow balls and 13 green balls. A ball is picked at random and not replaced. A second ball is then picked.

(a) Find the probability that the second ball is yellow.

(b) Given that exactly one ball is yellow, find the probability that it is the second one.

10.4 DISCRETE RANDOM VARIABLES

WORKED EXAMPLE 10.4

A discrete random variable Y has probability distribution as shown in the table below.

y	2	3	4	5
$P(Y = y)$	k	$\dfrac{2}{5}$	$\dfrac{1}{4}$	$2k$

(a) Find the exact value of k.

(b) Find the expected value of Y.

(a) $k + \dfrac{2}{5} + \dfrac{1}{4} + 2k = 1 \Rightarrow k = \dfrac{7}{60}$

> We use the fact that the probabilities must add up to 1.

(b) $E(Y) = 2\left(\dfrac{7}{60}\right) + 3\left(\dfrac{2}{5}\right) + 4\left(\dfrac{1}{4}\right) + 5\left(\dfrac{14}{60}\right) = 3.6$

> Find the expected value using the formula $E(Y) = \sum yP(Y = y)$.

Practice questions 10.4

10. Find the expected value of the following discrete random variable.

z	11	13	17	19	23
$P(Z = z)$	0.2	0.1	0.1	0.2	0.4

11. The random variable X has probability distribution as shown in the table below.

x	1	2	3	4
$P(X = x)$	c	p	0.4	0.2

Given that $E(X) = 2.6$, find the values of c and p.

12. A fair six-sided die has sides numbered $3k^2$ for $k = 1, 2, ..., 6$.
Find the expected score when the die is rolled once.

10.5 THE BINOMIAL DISTRIBUTION

WORKED EXAMPLE 10.5

When Sandra makes a sales call, the probability that the call is answered is 0.4.
She makes 15 calls every hour.

(a) Find the probability that during a particular hour, more than 10 of Sandra's calls are answered.

(b) Find the probability that more than 10 of Sandra's calls are answered during two out of seven working hours.

(a) Let X = number of calls answered during one hour.
Then $X \sim B(15, 0.4)$

> We start by defining the random variable and stating the probability distribution.
> There is a fixed number of calls and a fixed probability of success for each call, so we need to use the binomial distribution.

$$P(X > 10) = 1 - P(X \le 10)$$
$$= 1 - 0.99065\ldots \text{ (from GDC)}$$
$$= 0.00935 \text{ (3 SF)}$$

> Write down the probability required. To use the calculator we must relate it to $P(X \le k)$.

(b) Let Y = number of hours out of 7 in which more than 10 calls are answered. Then
$$Y \sim B(7, 0.00935)$$
$$P(Y = 2) = 0.00175 \text{ (3 SF) from GDC}$$

> There is a fixed number of hours, with a fixed probability of getting more than 10 calls answered in each hour, so we need to use the binomial distribution again. The probability of 'success' is the answer to part (a).

 When using your GDC, always state the distribution used and the probability calculated, and give the answer to 3 SF.

Practice questions 10.5

13. A fair six-sided die is rolled 16 times. Find the probability it lands on a '4' more than 5 times.

14. At the end of each training session, Anna takes 12 shots at a target. She knows that, on average, she can expect to hit the target 8 times.
   (a) State two conditions which are needed in order that the number of hits can be modelled by a binomial distribution.
   (b) Find the probability that Anna hits the target at least 7 times.

15. The random variable X has distribution $B(n, p)$. The mean of X is equal to three times its variance, and $P(X = 0) = 0.0123$ (to three significant figures). Find the values of n and p.

10.6 THE NORMAL AND INVERSE NORMAL DISTRIBUTIONS

WORKED EXAMPLE 10.6

Test scores are normally distributed with mean 62 and standard deviation 12.

(a) Find the percentage of candidates who scored below 50.

(b) It is known that 44% of candidates scored between 62 and p. Find the value of p.

(a) Let X = test score, so $X \sim N(62, 12^2)$

> We start by defining the random variable and stating the distribution used.

$P(X < 50) = 0.159$ (3 SF) from GDC
So 15.9% of candidates scored below 50.

> For a normal distribution you must be able to use a GDC to find a probability.

(b) $P(62 < X < p) = 0.44$

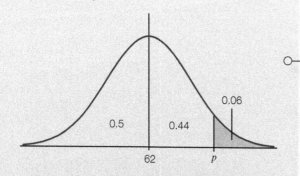

0.06

0.5   0.44

62   p

> Sketching a diagram allows us to see more clearly exactly what we need to find from the GDC.
>
> As the probability is known, we need to use the inverse normal distribution.

> On some calculators it is necessary to change problems like this into the form $P(X \le x) = \ldots$ (i.e. left-tail calculation).

From the diagram, $P(X > p) = 0.06$
From GDC, $p = 80.7$ (3 SF)

Practice questions 10.6

16. The random variable Y follows a normal distribution with mean 7.5 and variance 1.44. Find:

    (a) $P(6 < Y < 7)$

    (b) $P(Y \ge 8.5)$

    (c) the value of k such that $P(Y \le k) = 0.35$.

17. The weights, W kg, of babies born at a certain hospital satisfy $W \sim N(3.2, 0.7^2)$. Find the value of m such that 35% of the babies weigh between m kg and 3.2 kg.

18. The time a laptop battery can last before needing to be recharged is assumed to be normally distributed with mean 4 hours and standard deviation 20 minutes.

    (a) Find the probability that a laptop battery will last more than 4.5 hours.

    (b) A manufacturer wants to ensure that 95% of batteries will last for $4 \pm x$ hours. Find x.

10.7 USING THE STANDARD NORMAL DISTRIBUTION WHEN μ OR σ IS UNKNOWN

WORKED EXAMPLE 10.7

It is known that the average height of six-year-old boys is 105 cm and that 22% of the boys are taller than 110 cm. Find the standard deviation of the heights.

Let X = height of a six-year-old boy. Then $X \sim N(105, \sigma^2)$.

> We start by defining the random variable and stating the distribution.

The standardised variable is $Z = \dfrac{X - 105}{\sigma}$ and $Z \sim N(0,1)$.

> As σ is unknown, we need to use the standard normal distribution, $Z \sim N(0, 1)$ where $Z = \dfrac{X - \mu}{\sigma}$.

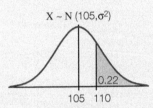 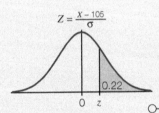

> ⚠ It is always a good idea to sketch a normal distribution diagram to help you visualise the solution.

> As the probability is known (22% = 0.22), we need to use the inverse normal distribution.

$P(Z > z) = 0.22$

$\Rightarrow z = 0.7722$ (from GDC)

> We can now find σ from z.

$z = \dfrac{x - 105}{\sigma}$

$\therefore 0.7722 = \dfrac{110 - 105}{\sigma}$

$\Rightarrow \sigma = 6.48 \text{ cm}$

Practice questions 10.7

19. The weights of apples sold at a market are normally distributed with mean weight 125 g. It is found that 26% of the apples weigh less than 116 g. Find the standard deviation of the weights.

20. A machine dispenses cups of coffee. The volume of coffee in a cup is normally distributed with standard deviation 5.6 ml. If 10% of cups contain more than 160 ml, find to one decimal place the mean volume of coffee in a cup.

21. The random variable X follows the distribution $N(\mu, \sigma^2)$, and $P(X \le 3.6) = P(X \ge 8.2) = 0.32$. Write down the value of μ and find the standard deviation of the distribution.

Mixed practice 10

1. The heights of trees in a forest are normally distributed with mean height 26.2 m and standard deviation 5.6 m.

   (a) Find the probability that a tree is more than 30 m tall.

   (b) What is the probability that among 16 randomly selected trees at least two are more than 30 m tall?

2. A discrete random variable X is given by $P(X = n) = kn^2$ for $n = 1, 2, 3, 4$.

   Find the value of k and the expected value of X.

3. If $P(A) = 0.3$, $P(B \mid A') = 0.5$ and $P(A \mid B) = \dfrac{7}{16}$, find $P(B \mid A)$.

4. The discrete random variable Y has probability distribution as shown in the table below.

y	1	2	3	4
$P(Y = y)$	0.1	0.2	p	q

   Given that $E(Y) = 3.1$, find the values of p and q.

5. A fair red die has sides labelled 1, 1, 1, 4, 5, 6. A fair blue die has sides labelled 1, 2, 3, 4, 5, 6. The two dice are rolled and the results are added. Find the probability that:

   (a) the total is greater than 7

   (b) the total is 10 given that it is not 7.

6. 30% of clothes sold by a shop are pink, and 60% are dresses. Given that the colour of the item sold is independent of whether or not it is a dress, find the proportion of clothes sold in the shop which are neither pink nor dresses.

7. A company hires out vans on a daily basis. It has three vans it can hire out. The number of requests it gets for hiring a van can be modelled by the following distribution:

Number of requests	0	1	2	3	4	5
Probability	0.07	0.22	0.35	0.21	0.12	0.03

   (a) Find the probability that in one day some requests have to be turned down.

   (b) Given that some requests have to be turned down, find the probability that there were exactly four requests.

   (c) Find the probability that in a seven-day week there are at least two days in which requests are rejected.

   (d) Find the probability distribution of the number of vans that are hired out each day.

   (e) The price of hiring a van is $120. Find the expected daily takings of the company.

   (f) The number of kilometres travelled by each van can be modelled by a normal distribution with mean 150 km. 10% of vans travel more than 200 km. Find the standard deviation of the normal distribution.

   (g) If two vans are hired, find the probability that each travels less than 100 km.

Going for the top 10

1. A large box contains three different types of toys. One third of the toys are cars, one quarter are yo-yos and the rest are balloons. 20% of the cars, 30% of the yo-yos and 40% of the balloons are pink. A toy is selected at random from the box. Given that the toy is pink, find the probability that it is a balloon.

2. It is known that the scores on a test follow a normal distribution $N(\mu, \sigma^2)$. 20% of the scores are above 82 and 10% are below 47.

   (a) Show that $\mu + 0.84162\sigma = 82$.

   (b) By writing a similar equation, find the mean and standard deviation of the scores.

3. A bag contains 12 red balls, 20 green balls and 18 blue balls.

   (a) If three balls are picked at random, find the probability that they are all the same colour.

   (b) If three balls are picked at random, find the probability that they are all different colours.

   (c) Given that the first two balls picked are blue, find the probability that the third ball picked is blue.

   (d) Find the probability that the third ball picked is red.

4. Three basketball players, Annie, Brent and Carlos, try to shoot a free throw. Annie shoots first, then Brent, then Carlos. The probability that Annie scores is 0.6, the probability that Brent scores is 0.5, and the probability that Carlos scores is 0.8. The shots are independent of each other and the first player to score wins.

   (a) Find the probability that Annie wins with her second shot.

   (b) What is the probability that Carlos gets a second shot?

   (c) (i) Show that the probability of Brent winning with his kth shot is $0.2 \times 0.04^{k-1}$.

   (ii) Hence find the probability that Brent wins.

11 EXAMINATION SUPPORT

COMMON ERRORS

There are several very common errors which you need to be aware of.

- Making up rules which don't exist, such as:

  - $\ln(x+y) = \ln x + \ln y$
    [*Note*: $\ln(x+y)$ cannot be simplified.]

  - $(x+y)^2 = x^2 + y^2$
    [*Note*: The correct expansion is $x^2 + 2xy + y^2$.]

  - $\dfrac{d}{dx}(f(x) \times g(x)) = \dfrac{d}{dx}f(x) \times \dfrac{d}{dx}g(x)$
    [*Note*: You cannot differentiate products by finding the derivatives of the factors and multiplying them together.]

- Algebraic errors, especially involving minus signs and brackets, such as:

  - $3 - (1 - 2x) = 3 - 1 - 2x$
    [*Note*: The correct expansion is $3 - 1 + 2x$.]

  - $(5x)^3 = 5x^3$
    [*Note*: The correct expansion is $125x^3$.]

- Arithmetic errors, especially involving fractions, such as:

  - $3 \times \dfrac{2}{5} = \dfrac{6}{15}$
    [*Note*: The correct answer is $\dfrac{6}{5}$.]

Spot the common errors

Find the errors in the solutions below.

1. Find $\int x^2 e^{2x}\,dx$.

Solution: $\int x^2 e^{2x}\,dx = \dfrac{x^3}{3} \times 2e^{2x}$

2. Solve $\ln x - \ln(10 - x) = \ln\left(\dfrac{x}{2}\right)$.

Solution: $\ln x - \ln 10 - \ln x = -\ln\left(\dfrac{x}{2}\right)$

$$-\ln 10 = \ln\left(\dfrac{x}{2}\right)$$

$$-10 = \dfrac{x}{2} \text{ so } x = -20$$

3. If $f(x) = \dfrac{1}{3-x}$, find $f \circ f(x)$.

Solution: $f \circ f(x) = \dfrac{1}{3 - \dfrac{1}{3-x}} = \dfrac{3-x}{3(3-x)-1} = \dfrac{1}{3-1} = \dfrac{1}{2}$

HOW TO CHECK YOUR ANSWERS

What you need to know

Checking questions by reading through your previous working is usually not very effective. You need to try more subtle methods such as:

- using your calculator to check a solution you obtained algebraically, and vice versa
- estimating the answer
- substituting numbers into algebraic expressions.

It is vital that you know how to use your calculator to check work. It often requires a little imagination.

Example 1: Definite integration

Suppose you were asked to find $\int xe^{2x}\,dx$ and you said that the answer was $e^{2x}(2x-1)+c$. To check this, turn it into a definite integral.

According to your answer:

$$\int_3^4 xe^{2x}\,dx = \left[e^{2x}(2x-1)\right]_3^4 = 7e^8 - 5e^6 \approx 18\,850$$

 The numbers 3 and 4 were randomly chosen. Most numbers would be suitable, but try to keep them simple while avoiding 0, 1 and any numbers given in the question – all of which might lead to errors not getting caught.

But according to the definite integral function on your calculator, $\int_3^4 xe^{2x}\,dx \approx 4712$, so you have made a mistake! This does not necessarily mean, however, that you have to go all the way back to the beginning. Inspecting the two numbers shows that your answer is exactly four times too big, so you have a hint as to what the correct answer should be!

Example 2: Algebraic manipulation

Suppose you were asked to expand $(1 + 2x)^4$ and you got $1 + 8x + 12x^2 + 8x^3 + 2x^4$.

Substituting $x = 1$ into the original expression gives $3^k = 81$. Substituting $x = 1$ into your answer gives 31, so something has gone wrong.

 This method (substitution) can be used to check each line of working to identify where the mistake occurred. It can also be used to check the steps when you are proving an identity.

Example 3: Differentiation

Suppose you were asked to differentiate $x^2 \sin x$ and you said that the answer was $2x \sin x + x^2 \cos x$.

If this is the derivative, then the gradient at $x = 3$ would be $6 \sin 3 + 9 \cos 3$, which is -8.06. According to the differentiation function on the calculator, the gradient at $x = 3$ is -8.06, so your answer is plausible.

 Just because these numbers agree doesn't mean the answer is correct – it could simply be a coincidence. However, it should certainly give you the confidence to move on and check another question!

 If you are finding the equation of a tangent, you should also plot it and the original curve on the same graph to make sure it looks like a tangent.

Example 4: Problems involving parameters

Many questions try to eliminate the option of using a calculator by putting an additional unknown into the question. For example, suppose a question asked you to find the range of the function $f(x) = x^2 - 2ax + 3a^2$, giving your answer in terms of a.

If you thought the answer was $f(x) \geq 2a^2$, then you could check this by sketching the graph with $a = 3$. The minimum point on this graph turns out to be 18, which is consistent with the range being $f(x) \geq 2a^2$.

12 THINGS YOU NEED TO KNOW HOW TO DO ON YOUR CALCULATOR

CASIO CALCULATOR

Note: These instructions were written based on the calculator model fx-9860G SD and may not be applicable to other models. If in doubt, refer to your calculator's manual.

Skill	On a Casio calculator	
Store numbers as variables	In the RUN menu use the $\rightarrow$ button and then key in a letter.	sin 60 0.8660254038 Ans→P 0.8660254038 P²+1 1.75 ▶MAT
Solve equations graphically	In the GRAPH menu input one equation into Y1 and the other into Y2. Go to the G-Solv menu (F5), and use the ISCT (F5) function to find the intersection. If there is more than one intersection, press across to move the cursor to the next intersection point.	Y1=1-0.4X Y2=sin X ISECT X=3.594937595 Y=-0.437975038
Solve equations numerically	In the RUN menu press OPTN, CALC (F4) and then SolveN (F5). Put in the equation you want to solve.	SolveN(cos X=tan (2X)) ▶MAT Ans 1 -7.853 2 -5.908 3 -4.712 4 -3.516 5 -1.57 -7.853981634

Skill	On a Casio calculator

Solve linear simultaneous equations	In the EQUA menu press F1 for simultaneous equations.
	
	Rearrange the equations into the form $ax + by + cz = d$ and input all the coefficients.
	
	`anX+bnY+CnZ=dn`
	`                a      b    c      d`
	`1[  1  ]   2    3    4]`
	`2[  1      3    5    7]`
	`3[  0      1   -1    0]`
	
	`                                1`
	`[SOLV][DEL][CLR][EDIT]`

Use modulus functions	In the RUN or GRAPH menu press OPTN then across (F6) and NUM (F4).
	
	Abs (F1) is the modulus function.
	
	`Abs (3-π)`
	`            0.1415926536`
	
	`[Abs][Int][Frac][Rnd][Intg][ ▷ ]`

Work with complex numbers	In the RUN menu, using SHIFT 0 allows you to enter 'i' and use it for calculations.
	
	By pressing OPTN, CPLX (F3) you can access a list of functions which can be applied to complex numbers, such as finding the modulus or the argument.
	
	You can enter complex numbers in polar form by using the $re^{i\theta}$ notation.
	
	`(1+i)÷(1-i)`
	`                        i`
	`Arg Ans`
	`            1.570796327`
	`e^(iπ)`
	`                       -1`
	`[ i ][Abs][Arg][Conj]   [ ▷ ]`

Put sequences into lists	Go to the STAT menu. With the cursor over List 1 press OPTN, List (F1) and then Seq (F5). The syntax is
	
	Seq(*expression*, X, *lowest value, highest value, increment*).
	
	To find the sum of the sequence, go to List 2 and, using the same menu as before, go across to get Cuml then List 1.
	
	`       List 1  List 2  List 3  List 4`
	`SUB`
	`1        2      2`
	`2        5      7`
	`3        8     15`
	`4       11     26`
	`Cuml List 1`

Skill	On a Casio calculator

Find numerical derivatives	In the RUN menu press OPTN, CALC (F4) and then d/dx (F2).
	
	Input the expression you want to differentiate followed by a comma and the value at which you want to evaluate the derivative.
	
	If you want to sketch a derivative you can use this expression in a graph too.
	
	`d/dx(sin (3X),π)`
	`                    -3`
	`▶MAT` 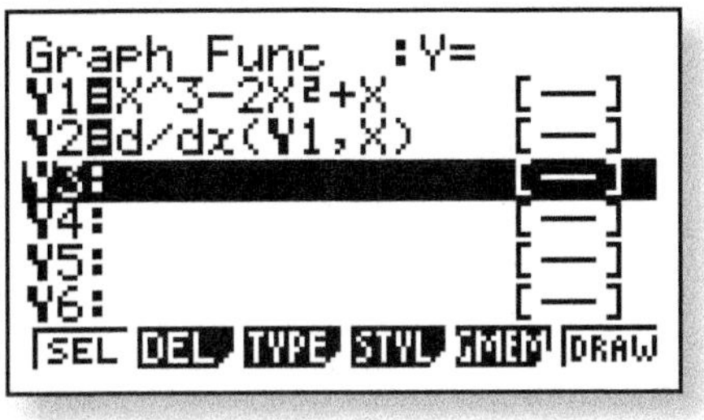
	
	`Graph Func   :Y=`
	`Y1▣X^3-2X²+X      [—]`
	`Y2▣d/dx(Y1,X)     [—]`
	`Y3:               [  ]`
	`Y4:               [—]`
	`Y5:               [—]`
	`Y6:               [—]`
	`[SEL] DEL▸ TYPE▸ STYL▸ GMEM▸ DRAW`

Find numerical integrals	In the RUN menu press OPTN, CALC (F4) and then $\int dx$ (F4). Input the expression you want to integrate followed by the lower and upper limits, all separated by commas.
	
	`∫((sin X)²,0,π)`
	`            1.570796327`
	
	`Solve d/dx d²/dx² ∫dx SolvN  ▷`

Find maximum and minimum points on a graph	In the GRAPH menu, plot the graph and press G-Solv (F5).
	
	Then press Max (F2) or Min (F3).
	
	`Y1=X^3-2X²+X`
	
	`                        MAX`
	`X=0.3333333319   Y=0.1481481481`

Find sample statistics	In the STAT menu enter the data in List 1 and, if required, the frequencies in List 2. Press CALC (F2). Use SET (F6) to make sure that the 1-Var XList is List 1 and that 1-Var Freq is either 1 or List 2, as appropriate. Then exit and press 1-Var (F1). Scrolling down shows the median and quartiles.
	
	`1-Variable`
	`x̄    =3.9090909`
	`Σx    =172`
	`Σx²   =860`
	`σx    =2.06505758`
	`sx    =2.08893187`
	`n     =44          ↓`

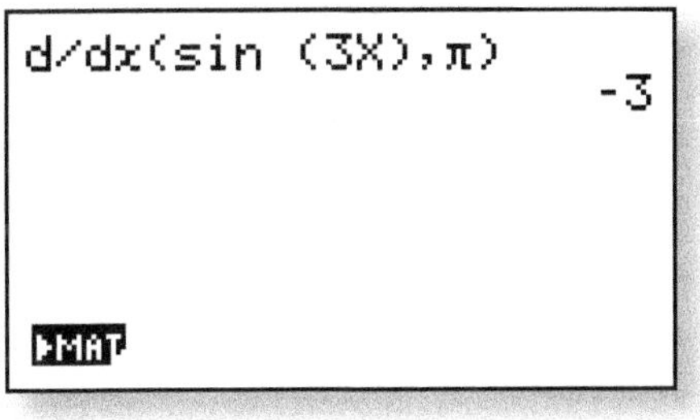

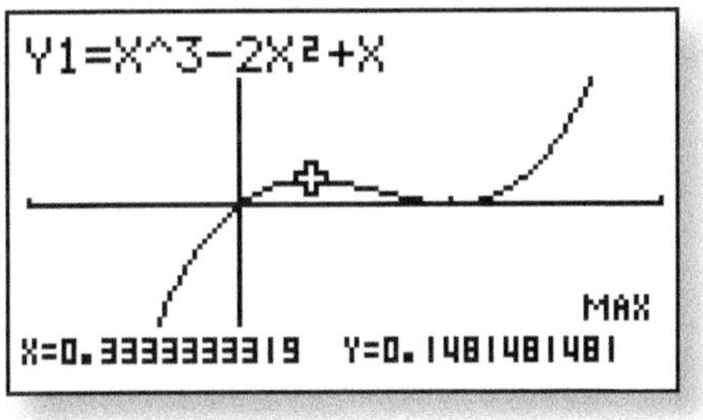

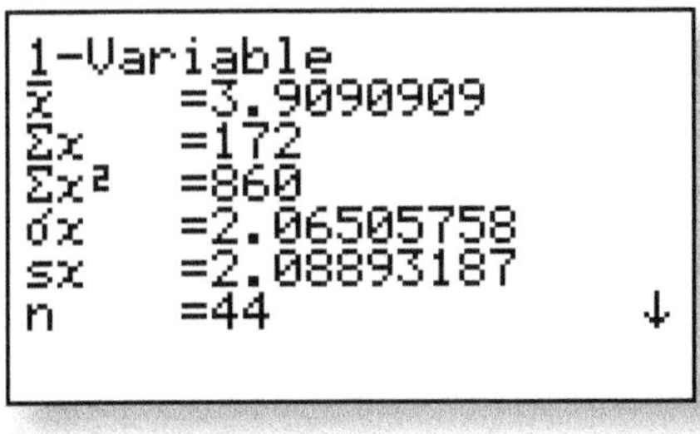

Skill	On a Casio calculator
Find probabilities in distributions	In the RUN menu, press OPTN, STAT (F5) and then DIST (F3). There are several options:

Name	Description	Syntax
Ncd	Probability from a normal $N(\mu, \sigma^2)$ distribution	NormalCD(*lower limit, upper limit*, μ, σ)
Bpd	Probability from a binomial $B(n, p)$ distribution	BinomialPD(x, n, p)
Bcd	Cumulative probability, $P(X \leq x)$, from a binomial $B(n, p)$ distribution	BinomialCD(x, n, p)
Ppd	Probability from a Poisson Po(m) distribution	PoissonPD(x, m)
Pcd	Cumulative probability, $P(X \leq x)$, from a Poisson Po(m) distribution	PoissonCD(x, m)

```
BinominalCD(1,3,0.5)
                  0.5
PoissonPD(2,2.4
        0.2612677055



Ppd  Pcd  InvP
```

Use inverse normal functions	To find the boundaries of a region with a specified probability, go to the STAT menu, then DIST (F5), NORM (F1) and InvN (F3). Input the data as a variable.

```
Inverse Normal
Data    :Variable
Tail    :Left
Area    :0.95
σ       :1
μ       :0
Save Res:None         ↓
```

Depending on the information you have, you can use different 'tails': for $P(X < x)$ use the left tail, for $P(X > x)$ use the right tail, and for $P(-x < X < x)$ use the central tail.

Input the mean and standard deviation. Use $\mu = 0$ and $\sigma = 1$ if you do not know the mean or standard deviation and want to find a Z-score.

TEXAS CALCULATOR

Note: These instructions were written based on the calculator model TI-84 Plus Silver Edition and may not be applicable to other models. If in doubt, refer to your calculator's manual.

Skill	On a Texas calculator	

Store numbers as variables	In the RUN menu, use the STO▶ button and then key in a letter.	sin(60)
		        .8660254038
		Ans→P
		        .8660254038
		P²+1
		              1.75

Solve equations graphically	In the Y= menu, input one equation into Y1 and the other into Y2. Then press the GRAPH button. You may need to use the WINDOW or ZOOM functions to find an appropriate scale. To look for intersections, press CALC (2nd, F4) then intersect (F5). You must select the two graphs you want to intersect and move the cursor close to the intersection point you are interested in.	Intersection X=3.5949376 Y=-.437975
Solve equations numerically	In the MATH menu press Solve (0). You must rearrange your equation into the form … = 0. Input this and press Solve (ALPHA, ENTER) when the cursor is above the X value. To find other values, change the bounds within which you want the calculator to search. For polynomial equations, you can find all solutions by using the solver in the PolySmlt 2 APP (which is recommended by the IB).	cos(X)−tan(2X)=0 •X=■37473443270… bound={-1ε99,1… •left-rt=0

Skill	On a Texas calculator
Solve linear simultaneous equations	In the PolySmlt APP select 'SimultEqnSolver' (2). You can change the number of equations and the number of unknowns. Rearrange the equations into the form $ax + by + cz = d$ and input all the coefficients. If the solution is not unique, a parametric representation of the solution will be given. SYSTEM MATRIX (3×4) [1 1 1 1] [1 2 3 4] [2 3 4 5] (3,4)=5 [MAIN][MODE][CLR][LOAD][SOLVE] SOLUTION SET $x_1 = -2+x_3$ $x_2 = 3-2x_3$ $x_3 = x_3$ [MAIN][MODE][SYSM][STO][RREF]
Use modulus functions	In the MATH menu, go across to NUM; the first option is ABS, which is the calculator notation for the modulus function. abs(3-π) .1415926536
Work with complex numbers	2nd, . (decimal point) allows you to enter 'i' and use it for calculations. By pressing MATH and going across to CPX you can access a list of functions which can be applied to complex numbers, such as finding the modulus (abs) or the argument (angle). You can enter complex numbers in polar form by using the $re^{i\theta}$ notation. (1+i)/(1-i) i angle(Ans) 1.570796327 e^(iπ) -1
Put sequences into lists	Press LIST (2nd, STAT) and move across to OPS. Option 5 is seq, an operation which puts a sequence into a list. The syntax is seq(*rule*, X, *lower limit*, *upper limit*, *step*). You can store this sequence in a list using the STO▶ button. To look at the cumulative sum of your sequence, use the cumSum function from the same menu. seq(3X-1,X,1,10,1)→L₁ {2 5 8 11 14 17… cumSum(L₁) {2 7 15 26 40 5…

Skill	On a Texas calculator	

Find numerical derivatives	In the MATH menu, option 8 is the numerical derivative, nDeriv. The syntax is nDeriv(*function*, X, *value of interest*).
	
	If you want to sketch the derivative function, you can graph Y1 = nDeriv(*function*, X, X).
		`nDeriv(sin(3X),X`
		`,π)`
		`          -2.9999955`

Find numerical integrals	In the MATH menu, option 9 is numerical integration, fnInt. The syntax is fnInt(*function*, X, *lower limit*, *upper limit*).
	
	If you want to see how the value of the integral changes with the upper limit, you can graph Y1 = fnInt(*function*, X, *lower limit*, X).
		`fnInt((sin(X))²,`
		`X,0,π)`
		`          1.570796327`

Find maximum and minimum points on a graph	When viewing a graph in the CALC menu (2nd, F4), press minimum (3) or maximum (4). Use the cursor to describe the left and right sides of the region you want to look in and then click the cursor close to the stationary point.	`Maximum` `X=.33333265   Y=.14814815`

Find sample statistics	In the STAT menu, use the edit function to enter the data in List 1 and, if required, the frequencies in List 2. Press STAT and CALC, then 1-Var Stats (1) as appropriate.
	
	Give the name of the list which holds the data and, if required, the list which holds the frequencies. Scrolling down shows the median and quartiles.
		`1-Var Stats`
		`x̄=9.868421053`
		`Σx=375`
		`Σx²=5161`
		`Sx=6.282412399`
		`σx=6.199197964`
		`↓n=38`

Skill	On a Texas calculator
Find probabilities in distributions	Probabilities from different distributions can be found using the DISTR menu (2nd, VARS). There are several options:

Name	Description	Syntax
normalcdf	Probability from a normal $N(\mu, \sigma^2)$ distribution	normalcdf(*lower limit, upper limit*, μ, σ)
binomialpdf	Probability from a binomial $B(n, p)$ distribution	binompdf(n, p, x)
binomialcdf	Cumulative probability, $P(X \le x)$, from a binomial $B(n, p)$ distribution	binomcdf(n, p, x)
poissonpdf	Probability from a Poisson $Po(m)$ distribution	poissonpdf(m, x)
poissoncdf	Cumulative probability, $P(X \le x)$, from a Poisson $Po(m)$ distribution	poissoncdf(m, x)

```
normalcdf(-1E99,
150,160,10)
         .1586552596
binompdf(12,0.4,
5)
         .2270303355
■
```

Use inverse normal functions	If you know the probability of an event being below a particular point of a normal distribution, you can find the value of that point. In the DISTR menu (2nd, VARS), use the invNorm function with the syntax invNorm(*probability*, μ, σ). Use $\mu = 0$ and $\sigma = 1$ if you do not know the mean or standard deviation and want to find a Z-score.

```
invNorm(0.6,0,1)
         .2533471011
```

13 WORKED SOLUTIONS

1 EXPONENTS AND LOGARITHMS

Mixed practice 1

1. $3 \times 9^x - 10 \times 3^x + 3 = 0$

$\Leftrightarrow 3 \times \left(3^2\right)^x - 10 \times 3^x + 3 = 0$

$\Leftrightarrow 3 \times \left(3^x\right)^2 - 10 \times 3^x + 3 = 0$

Let $3^x = y$. Then

$3y^2 - 10y + 3 = 0$

$\Leftrightarrow (3y - 1)(y - 3) = 0$

$\Leftrightarrow y = \dfrac{1}{3}$ or $y = 3$

So $3^x = \dfrac{1}{3} \Rightarrow x = -1$

or $3^x = 3 \Rightarrow x = 1$

2. $2^{3x+1} = 5^{5-x}$

$\Rightarrow \ln\left(2^{3x+1}\right) = \ln\left(5^{5-x}\right)$

$\Rightarrow (3x + 1)\ln 2 = (5 - x)\ln 5$

$\Rightarrow (3\ln 2)x + \ln 2 = 5\ln 5 - x\ln 5$

$\Rightarrow (3\ln 2)x + x\ln 5 = 5\ln 5 - \ln 2$

$\Rightarrow x(3\ln 2 + \ln 5) = 5\ln 5 - \ln 2$

$\Rightarrow x = \dfrac{5\ln 5 - \ln 2}{3\ln 2 + \ln 5}$

3. From the first equation:

$\ln\left(x^2 y\right) = 15$

$\Rightarrow x^2 y = e^{15}$

$\Rightarrow y = \dfrac{e^{15}}{x^2}$

Substituting this into the second equation gives

$\ln x + \ln\left(\dfrac{e^{15}}{x^2}\right)^3 = 10$

$\Rightarrow \ln\left(x \times \dfrac{e^{45}}{x^6}\right) = 10$

$\Rightarrow \dfrac{e^{45}}{x^5} = e^{10}$

$\Rightarrow x^5 = e^{35}$

$\Rightarrow x = e^7$

Then, substituting back to find y:

$y = \dfrac{e^{15}}{x^2} = \dfrac{e^{15}}{e^{14}} = e$

So the solution is $x = e^7$, $y = e$.

4. $y = \ln x - \ln(x + 2) + \ln\left(x^2 - 4\right)$

$= \ln\left(\dfrac{x}{x + 2}\right) + \ln\left(x^2 - 4\right)$

$= \ln\left(\dfrac{x\left(x^2 - 4\right)}{x + 2}\right)$

$= \ln\left(\dfrac{x(x - 2)(x + 2)}{x + 2}\right)$

$= \ln\left(x(x - 2)\right)$

$\therefore \ x(x - 2) = e^y$

$\Rightarrow x^2 - 2x - e^y = 0$

$\Rightarrow x = \dfrac{2 \pm \sqrt{4 + 4e^y}}{2} = 1 \pm \sqrt{1 + e^y}$

So $x = 1 + \sqrt{1 + e^y}$ (as $x > 0$)

5. Substituting $x = 5$ and $y = \ln 16$ into the equation:

$\ln 16 = 4\ln(5 - a)$

$\Rightarrow \ln 16 = \ln(5 - a)^4$

$\Rightarrow 16 = (5 - a)^4$

$\Rightarrow 5 - a = \pm 2$

$\Rightarrow a = 3$ or 7

Putting each of these values into the original equation and checking that $(5, \ln 16)$ is a solution shows that $a = 7$ is not valid, because in that case we would get $\ln(5 - 7) = \ln(-2)$, which is not real.

So $a = 3$.

6. (a) (i) After 10 days the rate of increase is 325 per day, so $\dfrac{dD}{dt} = 325$ when $t = 10$:

$\dfrac{dD}{dt} = 0.2Ce^{-0.2t}$

$325 = 0.2Ce^{-2}$

$\therefore C = 1625e^2$

(ii) After 10 days the demand is 15 000, so we have $D = 15\,000$ when $t = 10$:

$15\,000 = A - Ce^{-2}$

$15\,000 = A - \left(1625e^2\right)e^{-2}$

$\therefore A = 16\,625$

The initial demand, D_0, is the value of D when $t = 0$:

$$D_0 = A - C = 16\,625 - 1625e^2 = 4618 \text{ (to the nearest integer)}$$

(iii) As $t \to \infty$, $e^{-0.2t} \to 0$. Therefore $D \to A = 16625$.

(b) (i) We are given that $D = 16\,625 - 1625e^2$ when $t = 0$, so

$$16\,625 - 1625e^2 = B\ln\left(\frac{0+10}{5}\right)$$

$$\therefore B = \frac{16\,625 - 1625e^2}{\ln 2} = 6662 \text{ (4 SF)}$$

(ii) As $t \to \infty$, $\ln\left(\frac{t+10}{5}\right) \to \infty$. Therefore $D \to \infty$.

(c) We need to find the first t for which
$B\ln\left(\frac{t+10}{5}\right) > A - Ce^{-0.2t}$; that is, we need to

solve $B\ln\left(\frac{t+10}{5}\right) = A - Ce^{-0.2t}$. From GDC,

$t = 50.6$, i.e. after 51 days.

Going for the top 1

1. $2^{3x-4} \times 3^{2x-5} = 36^{x-2}$

$\Rightarrow \ln\left(2^{3x-4} \times 3^{2x-5}\right) = \ln\left(36^{x-2}\right)$

$\Rightarrow \ln\left(2^{3x-4}\right) + \ln\left(3^{2x-5}\right) = \ln\left(36^{x-2}\right)$

$\Rightarrow (3x-4)\ln 2 + (2x-5)\ln 3 = (x-2)\ln 36$

$\Rightarrow (3\ln 2)x - 4\ln 2 + (2\ln 3)x - 5\ln 3 = x\ln 36 - 2\ln 36$

$\Rightarrow (3\ln 2 + 2\ln 3 - \ln 36)x = 4\ln 2 + 5\ln 3 - 2\ln 36$

$\Rightarrow (\ln 8 + \ln 9 - \ln 36)x = \ln 2^4 + \ln 3^5 - \ln 36^2$

$\Rightarrow x\ln\left(\frac{8\times 9}{36}\right) = \ln\left(\frac{2^4 \times 3^5}{36^2}\right)$

$\Rightarrow x\ln 2 = \ln\left(\frac{6^4 \times 3}{6^4}\right)$

$\Rightarrow x = \frac{\ln 3}{\ln 2}$

2. Changing base a into base b, we have
$\log_a b = \frac{\log_b b}{\log_b a} = \frac{1}{\log_b a}$, so

$\log_a b^2 = c^2$

$\Leftrightarrow 2\log_a b = c^2$

$\Leftrightarrow 2\left(\frac{1}{\log_b a}\right) = c^2$

Then, substituting $\log_b a = c+1$ from the other equation gives

$$2\left(\frac{1}{c+1}\right) = c^2$$

$$\Leftrightarrow \frac{2}{c+1} = c^2$$

$$\Leftrightarrow 2 = c^3 + c^2$$

$$\Leftrightarrow c^3 + c^2 - 2 = 0$$

$$\Leftrightarrow (c-1)(c^2 + 2c + 2) = 0$$

$$\therefore c = 1$$

Therefore, $\log_b a = 1+1 = 2$ and hence $a = b^2$.

3. Gradient of the line through $(2, 4.5)$ and $(4, 7.2)$ is
$$m = \frac{7.2 - 4.5}{4-2} = 1.35$$

So

$$\ln F - 4.5 = 1.35(\ln x - 2)$$

$$\ln F = 1.35\ln x + 1.8$$

$$= \ln x^{1.35} + 1.8$$

and therefore $F = e^{\ln x^{1.35} + 1.8} = e^{1.8}x^{1.35}$.

2 POLYNOMIALS

Mixed practice 2

1. Using the quadratic formula:

$$x = \frac{-8 \pm \sqrt{8^2 - 4\times 1\times 3}}{2}$$

$$= \frac{-8 \pm \sqrt{52}}{2}$$

$$= \frac{-8 \pm 2\sqrt{13}}{2}$$

$$= -4 \pm \sqrt{13}$$

2. The x-coordinate of the vertex is half-way between the roots. So:

$$\frac{(a-4)+(a+2)}{2} = 5$$

$$\Leftrightarrow a - 1 = 5$$

$$\Leftrightarrow a = 6$$

3. (a) Using the binomial theorem:

$$(2-x)^5 = 2^5 + \binom{5}{1}2^4(-x) + \binom{5}{2}2^3(-x)^2 + \binom{5}{3}2^2(-x)^3 + \dots$$

$$= 32 - 80x + 80x^2 - 40x^3 + \dots$$

(b) Let $x = 0.01$ so that $(2 - x)^5 = 1.99^5$. Then

$$1.99^5 \approx 32 - 80 \times 0.01 + 80 \times 0.01^2 - 40 \times 0.01^3$$
$$= 31.20796$$

4. $y = x^2 + kx + 2$ touches the x-axis and so has one repeated root. Therefore

$$b^2 - 4ac = 0$$
$$k^2 - 4 \times 1 \times 2 = 0$$
$$k^2 = 8$$
$$k = \pm\sqrt{8} = \pm 2\sqrt{2}$$

5. (a) Completing the square:

$$x^2 - 8x + 25 = (x - 4)^2 - 16 + 25$$
$$= (x - 4)^2 + 9$$

(b) The expression $\dfrac{3}{x^2 - 8x + 25}$ has maximum value when the denominator has minimum value, and this occurs at the vertex of the parabola $y = x^2 - 8x + 25$. From the completed square form, the vertex is at $(4, 9)$, i.e. the minimum value of $x^2 - 8x + 25$ is 9.

Hence the maximum value of $\dfrac{3}{x^2 - 8x + 25}$ is $\dfrac{3}{9} = \dfrac{1}{3}$.

6. The general term of this binomial expansion is

$$\binom{8}{r}(x^3)^{8-r}\left(\frac{3}{x}\right)^r = \binom{8}{r}x^{24-3r}x^{-r}3^r = \binom{8}{r}x^{24-4r}3^r$$

The term independent of x will have power 0 for x; that is, $24 - 4r = 0$, so $r = 6$. Therefore, the term is

$$\binom{8}{6}(x^3)^2\left(\frac{3}{x}\right)^6 = 28x^6\left(\frac{729}{x^6}\right) = 20\,412.$$

7. The equation $3x - 2x^2 = k$ can be rewritten as $2x^2 - 3x + k = 0$. No solutions means that the discriminant is negative. So:

$$b^2 - 4ac < 0$$
$$(-3)^2 - 4 \times 2 \times k < 0$$
$$\Leftrightarrow 9 - 8k < 0$$
$$\Leftrightarrow 8k > 9$$
$$\Leftrightarrow k > \frac{9}{8}$$

8. $x^4 + x^2 - 20 = 0$
$$\Leftrightarrow (x^2 + 5)(x^2 - 4) = 0$$
$$\Leftrightarrow x^2 = -5 \text{ or } 4$$

But $x^2 \geq 0$, so we must have $x^2 = 4$.

Hence $x = \pm 2$.

9. (a) $ax - x^2 = x(a - x)$

So the solutions to $ax - x^2 = 0$ are $x = 0$ and $x = a$.

(b) $a = 4$

(c) The maximum height (vertex of the parabolic path) occurs half-way between the two roots, i.e. at $x = 2$. When $x = 2$, the height is $y = 4x - x^2 = 4$ m.

10. (a) $x^2 + 4x - 12 = 6x - 9$
$$\Leftrightarrow x^2 - 2x - 3 = 0$$
$$\Leftrightarrow (x - 3)(x + 1) = 0$$
$$\Leftrightarrow x = 3 \text{ or } -1$$

When $x = 3$, $y = 6 \times 3 - 9 = 9$

When $x = -1$, $y = 6 \times (-1) - 9 = -15$

So the coordinates are $(3, 9)$ and $(-1, -15)$.

(b) $x^2 + 4x - 12 = 0$
$$\Leftrightarrow (x + 6)(x - 2) = 0$$
$$\Leftrightarrow x = -6 \text{ or } 2$$

So the x-intercepts are $(-6, 0)$ and $(2, 0)$.

(c) Since the second graph also has x-intercepts $(-6, 0)$ and $(2, 0)$, its equation is $y = a(x + 6)(x - 2)$.

When $x = 0$, $y = 36$, so $36 = a(6)(-2)$, hence $a = -3$.

$$-3(x + 6)(x - 2) = -3(x^2 + 4x - 12)$$
$$= -3x^2 - 12x + 36$$

So $a = -3$, $b = -12$, $c = 36$.

(d) The x-coordinate of the vertex of both graphs is $\dfrac{-6 + 2}{2} = -2$.

The vertex of the first graph has y-coordinate
$$y = (-2)^2 + 4(-2) - 12 = -16$$

The vertex of the second graph has y-coordinate
$$y = -3(-2)^2 - 12(-2) + 36 = 48$$

So the distance between the vertices is $48 - (-16) = 64$.

1. Using the binomial theorem:

$$(2+x)(3-2x)^5$$

$$=(2+x)\left(3^5+\binom{5}{1}3^4(-2x)+\binom{5}{2}3^3(-2x)^2+...\right)$$

$$=(2+x)(243-810x+1080x^2+...)$$

The quadratic term will be

$$2\times1080x^2+x\times(-810x)=2160x^2-810x^2=1350x^2$$

2. A quadratic function has equation $y=a(x-p)(x-q)$, where p and q are the roots (x-intercepts of the graph).

From the graph, there are roots -2 and 8, so
$y=a(x+2)(x-8)$

When $x=3$, $y=32$, so $a(3+2)(3-8)=32$

$$\Rightarrow-25a=32$$

$$\Rightarrow a=-\frac{32}{25}=-1.28$$

Therefore, the equation is $y=-1.28(x+2)(x-8)$

$$=-1.28x^2+7.68x+20.48$$

3. $(1+ax)^n=1+nax+\dfrac{n(n-1)}{2}(ax)^2+...$

$$=1+nax+\dfrac{n(n-1)}{2}a^2x^2+...$$

So we have

$$\frac{n(n-1)}{2}a^2=54 \quad\cdots\;(1)$$

$$na=12 \quad\cdots\;(2)$$

From (2), $a=\dfrac{12}{n}$. Substituting this into (1) gives

$$\frac{n(n-1)}{2}\left(\frac{12}{n}\right)^2=54$$

$$\Leftrightarrow\frac{72n^2-72n}{n^2}=54$$

$$\Leftrightarrow 72n^2-72n=54n^2$$

$$\Leftrightarrow 18n(n-4)=0$$

$$\therefore\; n=4$$

Then, substituting back into (2) gives $a=3$.

4. Two solutions means that the discriminant is positive:

$$b^2-4ac>0$$

$$\Leftrightarrow 9^2-4k\times k>0$$

$$\Leftrightarrow 81-4k^2>0$$

$$\Leftrightarrow 4k^2<81$$

$$\Leftrightarrow k^2<\frac{81}{4}$$

k can be positive or negative, but it should not be 0 (or else the equation will no longer be quadratic),

so $-\dfrac{9}{2}<k<0$ or $0<k<\dfrac{9}{2}$.

3 FUNCTIONS, GRAPHS AND EQUATIONS

Mixed practice 3

1. (a)

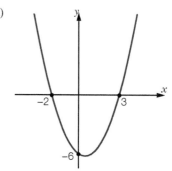

(b) $x<-2$ or $x>3$

2. To avoid taking the square root of a negative number, require:

$$x^2-a^2\geq0$$

$$\Rightarrow x\geq|a|\;\text{ or }\;x\leq-|a|$$

3. The graph has a half-period of $\dfrac{7\pi}{8}-\dfrac{3\pi}{8}=\dfrac{\pi}{2}$, while $y=\cos x$ has half-period of π. Therefore $y=\cos x$ has been stretched horizontally by a factor of $\dfrac{1}{2}$, and hence $b=2$.

Substituting $\left(\dfrac{3\pi}{8},0\right)$ into the equation
$y=a\cos(2x+c)$:

$$0=a\cos\left(2\times\frac{3\pi}{8}+c\right)$$

$$\Rightarrow\cos\left(\frac{3\pi}{4}+c\right)=0$$

$$\therefore\;\frac{3\pi}{4}+c=\frac{\pi}{2}$$

$$\Rightarrow c=-\frac{\pi}{4}$$

Substituting $\left(0, \dfrac{3}{\sqrt{2}}\right)$ into the equation

$y = a\cos\left(2x - \dfrac{\pi}{4}\right)$:

$\dfrac{3}{\sqrt{2}} = a\cos\left(2\times 0 - \dfrac{\pi}{4}\right)$

$\phantom{\dfrac{3}{\sqrt{2}}} = a\cos\left(-\dfrac{\pi}{4}\right)$

$\phantom{\dfrac{3}{\sqrt{2}}} = \dfrac{a}{\sqrt{2}}$

$\therefore a = 3$

Therefore $a = 3$, $b = 2$, $c = -\dfrac{\pi}{4}$.

4. (a)

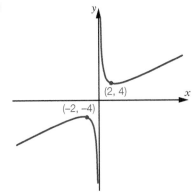

(b)

$y = 10$

(2, 4)

(−2, −4) 0.417   9.58

From GDC, when $y = 10$, $x = 0.417$ or 9.58

(c)

From the graph, the equation has one solution when k is the y-coordinate of a stationary point; that is, when $k = -4$ or 4.

5. (a)

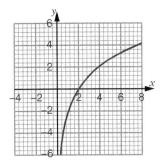

(b)

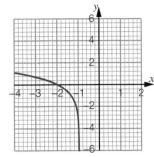

6. (a) When $m = 0$, $T = 5$, so $5 = c - k \times 1.2^0$

$\Rightarrow c - k = 5 \qquad \cdots (1)$

When $m = 5$, $T = 15$, so $15 = c - k \times 1.2^{-5}$

$\Rightarrow c - 1.2^{-5}k = 15 \qquad \cdots (2)$

Solving (1) and (2) simultaneously with a GDC gives $c = 21.7$, $k = 16.7$.

(b) Substituting in the values for c and k, we have the equation $T = 21.7 - 16.7 \times 1.2^{-m}$.

$T = 20 \Rightarrow 21.7 - 16.7 \times 1.2^{-m} = 20$.

Solving this equation with the GDC gives $m = 12.5$, i.e. it takes 12.5 minutes.

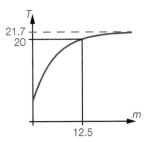

(c) Asymptote is $T = 21.7$. This means that the temperature of the room (which the butter approaches in the long run) is $21.7°C$.

7. $g(x) = x^3 - ax + b \Rightarrow g'(x) = 3x^2 - a$

$g(2) = -2 \Rightarrow 2^3 - 2a + b = -2$
$\qquad\qquad\quad \Rightarrow 2a - b = 10$
$g'(2) = 3 \Rightarrow 3 \times 2^2 - a = 3$
$\qquad\qquad\quad \Rightarrow a = 9$

Substituting $a = 9$ into the first equation gives
$18 - b = 10$, so $b = 8$.

8. (a)

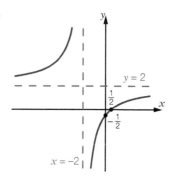

(b) $y = \dfrac{2x - 1}{x + 2}$

$\Rightarrow y(x + 2) = 2x - 1$

$\Rightarrow xy + 2y = 2x - 1$

$\Rightarrow 2x - xy = 2y + 1$

$\Rightarrow x(2 - y) = 2y + 1$

$\Rightarrow x = \dfrac{2y + 1}{2 - y}$

$\therefore f^{-1}(x) = \dfrac{2x + 1}{2 - x}$

Range is $y \in \mathbb{R}, y \ne -2$.

(c) $g(x) = f^{-1}(2 - x)$

$\quad = \dfrac{2(2 - x) + 1}{2 - (2 - x)}$

$\quad = \dfrac{5 - 2x}{x}$

Therefore, the domain is $x \in \mathbb{R}, x \ne 0$.

(d) $f(x) = g(x)$

$\dfrac{2x - 1}{x + 2} = \dfrac{5 - 2x}{x}$

$\Leftrightarrow x(2x - 1) = (5 - 2x)(x + 2)$

$\Leftrightarrow 2x^2 - x = 10 + x - 2x^2$

$\Leftrightarrow 4x^2 - 2x - 10 = 0$

$\Leftrightarrow 2x^2 - x - 5 = 0$

$\Leftrightarrow x = \dfrac{1 \pm \sqrt{1 - 4 \times 2 \times (-5)}}{2 \times 2}$

$\quad = \dfrac{1 \pm \sqrt{41}}{4}$

Going for the top 3

1. (a)

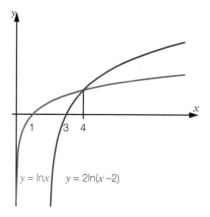

(b) $\ln x = 2\ln(x - 2)$

$\ln x = \ln(x - 2)^2$

$\Rightarrow x = x^2 - 4x + 4$

$\Rightarrow x^2 - 5x + 4 = 0$

$\Rightarrow (x - 1)(x - 4) = 0$

$x = 1$ or $x = 4$

But we need $x > 2$ for $\ln(x - 2)$ to be valid, so $x = 4$.

From the graph we can also see that there is only one intersection: $x = 4$.

2. Starting from $y = 2x^3 - 5x$:

Translation by 3 units in positive vertical direction gives $y = 2x^3 - 5x + 3$.

Translation by 2 units to the left gives
$y = 2(x + 2)^3 - 5(x + 2) + 3$.

Reflection in the x-axis gives
$y = -\left[2(x + 2)^3 - 5(x + 2) + 3\right]$

Expanding:

$y = -\left[2(x + 2)^3 - 5(x + 2) + 3\right]$

$\quad = -\left[2(x^3 + 6x^2 + 12x + 8) - 5(x + 2) + 3\right]$

$\quad = -\left[2x^3 + 12x^2 + 24x + 16 - 5x - 10 + 3\right]$

$\quad = -2x^3 - 12x^2 - 19x - 9$

So the resulting equation is $y = -2x^3 - 12x^2 - 19x - 9$

3. (a) $f(x) = \dfrac{ax - a^2 + 1}{x - a}$

$\quad = \dfrac{a(x - a) + 1}{x - a}$

$\quad = \dfrac{a(x - a)}{x - a} + \dfrac{1}{x - a}$

$\quad = a + \dfrac{1}{x - a}$

So $p = a$ and $q = 1$.

(b) The graph $y = \dfrac{1}{x}$ has been translated up by a and to the right by a.

(c)

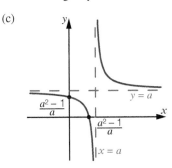

(d) Domain: $x \in \mathbb{R}, x \neq a$. Range: $f(x) \in \mathbb{R}, f(x) \neq a$

(e) $f(f(x)) = f\left(a + \dfrac{1}{x-a}\right)$

$= a + \dfrac{1}{\left(a + \frac{1}{x-a}\right) - a}$

$= a + \dfrac{1}{\frac{1}{x-a}}$

$= a + x - a$

$= x$

(f) Since $f \circ f(x) = x$, f is its own inverse and so $f^{-1}(x) = a + \dfrac{1}{x-a}$.

(g) As $f(x)$ and its inverse are the same, the graph is symmetrical about the line $y = x$.

4 SEQUENCES AND SERIES

Mixed practice 4

1. $u_4 = 17 \Rightarrow u_1 + 3d = 17 \qquad \cdots (1)$

$S_{20} = 990 \Rightarrow \dfrac{20}{2}[2u_1 + 19d] = 990$

$\qquad\qquad \Rightarrow 2u_1 + 19d = 99 \quad \cdots (2)$

From (1), $u_1 = 17 - 3d$; substituting this into (2) gives

$2(17 - 3d) + 19d = 99$

$\Leftrightarrow 34 + 13d = 99$

$\Leftrightarrow d = 5$

So $u_1 = 17 - 3d = 17 - 3 \times 5 = 2$.

2. The fourth, tenth and thirteenth terms of the geometric sequence are:

$u_4 = u_1 r^3$

$u_{10} = u_1 r^9$

$u_{13} = u_1 r^{12}$

As these form an arithmetic sequence:

$u_{10} - u_4 = u_{13} - u_{10}$

$\Rightarrow u_1 r^9 - u_1 r^3 = u_1 r^{12} - u_1 r^9$

$\Rightarrow r^{12} - 2r^9 + r^3 = 0 \quad (\text{as } u_1 \neq 0)$

$\Rightarrow r^9 - 2r^6 + 1 = 0 \quad (\text{as } r \neq 0)$

$\Rightarrow r = 1, \ 1.17, \ -0.852 \quad (\text{from GDC to 3 SF})$

$\therefore r = -0.852 \quad (\text{for sum to infinity to exist, } |r| < 1)$

3. $\displaystyle\sum_{r=1}^{12} 4r + \left(\dfrac{1}{3}\right)^r = 4\sum_{r=1}^{12} r + \sum_{r=1}^{12}\left(\dfrac{1}{3}\right)^r$

The first sum is an arithmetic series, and the second sum is a geometric series. So, using the formulae:

$4\left[\dfrac{12}{2}(2 \times 1 + (12-1) \times 1)\right] + \dfrac{\frac{1}{3}\left[1 - \left(\frac{1}{3}\right)^{12}\right]}{1 - \frac{1}{3}} = 312.5 \ (4 \text{ SF})$

4. For 20 terms of this series, i.e. with $n = 20$:

$\ln x + \ln x^4 + \ln x^7 + \ln x^{10} + \ldots$

$= \ln x + 4 \ln x + 7 \ln x + 10 \ln x + \ldots$

$= \ln x (1 + 4 + 7 + 10 + \ldots)$

$= \ln x \left(\dfrac{20}{2}(2 + 19 \times 3)\right)$

$= 590 \ln x$

$= \ln x^{590}$

5. We know that the total length of the pieces is 300, i.e. $S_n = 300$:

$S_n = \dfrac{n}{2}(u_1 + u_n)$

$300 = \dfrac{n}{2}(1 + 19)$

$\Leftrightarrow 300 = 10n$

$\Leftrightarrow n = 30$

So $u_{30} = 19$:

$u_n = u_1 + (n-1)d$

$19 = 1 + 29d$

$\Leftrightarrow d = \dfrac{18}{29}$ metres

6. The amount Aaron has in his account for the first few months is:

1st: 100

2nd: $100 + 110$

3rd: $100 + 110 + 120$

His monthly balance forms an arithmetic series with $a = 100$ and $d = 10$. So after n months he will have:

$$S_n = \frac{n}{2}\left[2 \times 100 + (n-1)10\right]$$

$$= \frac{n}{2}\left[200 + 10n - 10\right]$$

$$= \frac{n}{2}\left[190 + 10n\right]$$

$$= 5n^2 + 95n$$

The amount Blake has in his account for the first few months is:

1st: 50

2nd: 50×1.05

3rd: 50×1.05^2

His monthly balance forms a geometric series with $a = 50$ and $r = 1.05$. So after n months he will have:

$$S_n = \frac{50(1.05^n - 1)}{1.05 - 1}$$

$$= 1000(1.05^n - 1)$$

Therefore, Blake will have more in his account than Aaron does when:

$$1000(1.05^n - 1) > 5n^2 + 95n$$

$$\therefore n = 73 \text{ months (from GDC)}$$

Going for the top 4

1. (a) (i) Writing out the sum from the first term, a, to the nth term, ar^{n-1}:

$$S_n = a + ar + ar^2 + \ldots + ar^{n-2} + ar^{n-1} \quad (1)$$

Multiplying through by r:

$$rS_n = ar + ar^2 + ar^3 + \ldots + ar^{n-1} + ar^n \quad (2)$$

We can see that (1) and (2) have many terms in common:

$$S_n = a + ar + ar^2 + ar^3 + \ldots + ar^{n-2} + ar^{n-1} \quad (1)$$
$$rS_n = \quad\ \ ar + ar^2 + ar^3 + \ldots + ar^{n-2} + ar^{n-1} + ar^n \quad (2)$$

So $(2) - (1)$ gives:

$$rS_n - S_n = ar^n - a$$

$$\Rightarrow S_n(r-1) = a(r^n - 1)$$

$$\Rightarrow S_n = \frac{a(r^n - 1)}{r - 1}$$

(ii) If $-1 < r < 1$ (or equivalently $|r| < 1$), then as $n \to \infty$, $r^n \to 0$. So

$$S_n \to \frac{a(0-1)}{r-1} = \frac{-a}{r-1} = \frac{a}{1-r}$$

If $|r| \geq 1$, then r^n has no limit as $n \to \infty$, and so there is no finite limit for S_n.

Thus, $S_\infty = \dfrac{a}{1-r}$ only for $|r| < 1$.

(b) The sum of the first n terms is $S_n = \dfrac{u_1(1 - r^n)}{1 - r}$.

The sum of the next n terms is given by

$$S_{2n} - S_n = \frac{u_1(1 - r^{2n})}{1 - r} - \frac{u_1(1 - r^n)}{1 - r}$$

$$= \frac{u_1(1 - r^{2n} - 1 + r^n)}{1 - r}$$

$$= \frac{u_1(r^n - r^{2n})}{1 - r}$$

$$= \frac{u_1 r^n(1 - r^n)}{1 - r}$$

So the ratio of the sum of the first n terms to the sum of the next n terms is

$$\frac{u_1(1 - r^n)}{1 - r} : \frac{u_1 r^n(1 - r^n)}{1 - r} = 1 : r^n$$

(c) (i) $u_7 + 4u_5 = u_8$

i.e. $u_1 r^6 + 4u_1 r^4 = u_1 r^7$

$\Rightarrow r^2 + 4 = r^3$ (since $u_1, r \neq 0$)

$\Rightarrow r^3 - r^2 - 4 = 0$

To factorise this, substitute small positive and negative integers into $f(r) = r^3 - r^2 - 4$ until you find an r such that $f(r) = 0$:

$f(1) = 1^3 - 1^2 - 4 = -4$

$f(-1) = (-1)^3 - (-1)^2 - 4 = -6$

$f(2) = 2^3 - 2^2 - 4 = 0$

Therefore, by the factor theorem, $(r - 2)$ is a factor, and so by long division or equating coefficients we get:

$$r^3 - r^2 - 4 = (r - 2)(r^2 + r + 2)$$

For the quadratic $r^2 + r + 2$,

$\Delta = 1^2 - 4 \times 1 \times 2 = -7 < 0$

Hence there are no real roots for this quadratic factor, and so the only root of $f(r)$ is $r = 2$.

Then, from part (b), the ratio is

$1 : r^{10} = 1 : 2^{10} = 1 : 1024$

(ii) No, because $r > 1$ in this case, and as shown in part (a)(ii), the condition for a sum to infinity to exist is that $|r| < 1$.

2. The integers from 1 to 1000 form an arithmetic sequence with $u_1 = 1$, $u_n = 1000$ and $n = 1000$. So

$$S_n = \frac{n}{2}(u_1 + u_n)$$
$$= \frac{1000}{2}(1 + 1000)$$
$$= 500\,500$$

The multiples of 7 between 1 and 1000 form an arithmetic sequence with $u_1 = 7$, $u_n = 994$ and $n = \frac{994}{7} = 142$. So

$$S_n = \frac{n}{2}(u_1 + u_n)$$
$$= \frac{142}{2}(7 + 994)$$
$$= 71\,071$$

Therefore the sum of the integers between 1 and 1000 that are not divisible by 7 is

$500\,500 - 71\,071 = 429\,429$.

5 TRIGONOMETRY

Mixed practice 5

1. Using the formula for the area of a triangle:

$A = \frac{1}{2}ab\sin C$

$12 = \frac{1}{2}x(x+2)\sin 30°$

$\Rightarrow 12 = \frac{1}{2}x(x+2)\frac{1}{2}$

$\Rightarrow 48 = x^2 + 2x$

$\Rightarrow x^2 + 2x - 48 = 0$

$\Rightarrow (x+8)(x-6) = 0$

$\therefore x = 6$ (as $x > 0$)

2. By the sine rule:

$$\frac{\sin\theta}{5} = \frac{\sin 2\theta}{7}$$

$\Rightarrow 7\sin\theta = 5\sin 2\theta$

$\Rightarrow 7\sin\theta = 5(2\sin\theta\cos\theta)$

$\Rightarrow 7\sin\theta - 10\sin\theta\cos\theta = 0$

$\Rightarrow \sin\theta(7 - 10\cos\theta) = 0$

$\Rightarrow \sin\theta = 0$ or $\cos\theta = \dfrac{7}{10}$

$\Rightarrow \theta = 0°, 180°$ or $\theta = 45.6°$

$\therefore \theta = 45.6°$

3. (a) The depth of the water is 14 m when

$$d = 14 - 1.2\cos\left(\frac{\pi t}{12}\right) = 14,$$

which means that $\cos\left(\dfrac{\pi t}{12}\right) = 0$. This occurs

when $\dfrac{\pi t}{12} = \dfrac{\pi}{2}, \dfrac{3\pi}{2}, \dfrac{5\pi}{2}, \ldots$

The first positive value of t is $t = \dfrac{\pi}{2} \div \dfrac{\pi}{12} = 6$.

So the first time after midnight at which the depth is 14 m is 6:00 a.m.

(b) The smallest possible depth occurs when $\cos\left(\dfrac{\pi t}{12}\right)$ is at its maximum value of 1. When this happens, we have $d = 14 - 1.2 = 12.8$ m.

(c) The depth of water is less than 13.5 m when

$$14 - 1.2\cos\left(\frac{\pi t}{12}\right) < 13.5$$

$$1.2\cos\left(\frac{\pi t}{12}\right) > 0.5$$

$$\cos\left(\frac{\pi t}{12}\right) > 0.4166\ldots$$

Now, $\cos^{-1}(0.4166\ldots) = 1.141\ldots$

and $t \in [0, 24] \Rightarrow \dfrac{\pi t}{12} \in [0, 2\pi]$

So $\cos\left(\dfrac{\pi t}{12}\right) = 0.4166\ldots$ when

$\dfrac{\pi t}{12} = 1.141$ or $2\pi - 1.14 = 5.14$

Looking at the graph of $y = \cos\left(\dfrac{\pi t}{12}\right)$:

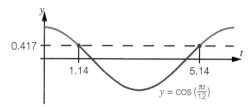

$y = \cos\left(\frac{\pi t}{12}\right)$

For $\dfrac{\pi t}{12} \in [0, 2\pi]$, $\cos\dfrac{\pi t}{12} > 0.4166...$ when

$$0 < \frac{\pi t}{12} < 1.14 \quad \text{or} \quad 5.14 < \frac{\pi t}{12} < 2\pi$$

corresponding to $t < 4.36$ (4:22 a.m.) and $t > 19.63$ (7:38 p.m.)

Alternatively, you could sketch the original graph, $y = 14 - 1.2\cos\left(\dfrac{\pi t}{12}\right)$, and answer the whole question using a GDC:

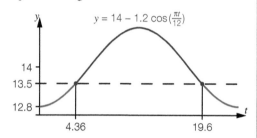

$y = 14 - 1.2\cos\left(\frac{\pi t}{12}\right)$

From the graph, the depth is less than 13.5 m for $t < 4.36$ or $t > 19.63$, i.e. before 4:22 a.m. and after 7:38 p.m.

4. $3\sin 2\theta = \tan 2\theta$

$\Leftrightarrow 3\sin 2\theta = \dfrac{\sin 2\theta}{\cos 2\theta}$

$\Leftrightarrow 3\sin 2\theta \cos 2\theta = \sin 2\theta$

$\Leftrightarrow 3\sin 2\theta \cos 2\theta - \sin 2\theta = 0$

$\Leftrightarrow \sin 2\theta(3\cos 2\theta - 1) = 0$

$\therefore \sin 2\theta = 0 \quad \text{or} \quad \cos 2\theta = \dfrac{1}{3}$

$\theta \in \left[-\dfrac{\pi}{2}, \dfrac{\pi}{2}\right] \Rightarrow 2\theta \in [-\pi, \pi]$

For $\sin 2\theta = 0$, the graph shows there to be three solutions in $[-\pi, \pi]$:

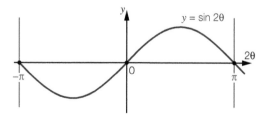

$y = \sin 2\theta$

One solution is $\sin^{-1} 0 = 0$.

Another is $\pi - 0 = \pi$.

Then, adding/subtracting 2π gives one further value in the interval: $\pi - 2\pi = -\pi$.

$\therefore 2\theta = 0, \pm \pi$

$\Rightarrow \theta = 0, \pm\dfrac{\pi}{2}$

For $\cos 2\theta = \dfrac{1}{3}$, the graph shows there to be two solutions:

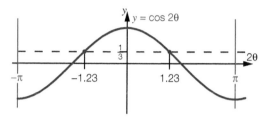

$y = \cos 2\theta$

One solution is $\cos^{-1}\left(\dfrac{1}{3}\right) = 1.23$.

$2\pi - 1.23 = 5.05$ is not in the interval $[-\pi, \pi]$, but $5.05 - 2\pi = -1.23$ is in the interval.

$\therefore 2\theta = \pm 1.23$

$\Rightarrow \theta = \pm 0.615$

So the solutions are $\theta = 0, \pm 0.615, \pm\dfrac{\pi}{2}$.

5. (a)   Since $O\hat{T}P$ is a right angle,

$$\text{Area} = \frac{1}{2}bh$$

$$= \frac{1}{2} \times 12 \times 7$$

$$= 42 \text{ cm}^2$$

(b)   Let $P\hat{O}T = \theta$. Then

$$\tan\theta = \frac{12}{7}$$

$$\therefore \theta = \tan^{-1}\left(\frac{12}{7}\right) = 1.04 \text{ (3 SF)}$$

(c)   Shaded area = area of triangle OPT
              − area of sector

$$= 42 - \frac{1}{2} \times 7^2 \times P\hat{O}T$$

$$= 16.5 \text{ (3 SF)}$$

6. **(a)** In triangle ABG, we know
$G\hat{A}B = 65°$ and $G\hat{B}A = 80°$, so

$$A\hat{G}B = 180° - 65° - 80° = 35°$$

By the sine rule:

$$\frac{AG}{\sin 80°} = \frac{20}{\sin 35°}$$

$$\Rightarrow AG = \frac{20}{\sin 35°} \times \sin 80°$$

$$= 34.3 \text{ m}$$

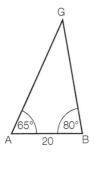

(b) In triangle AGT, we know the length AG and the angle $G\hat{A}T$:

$$\tan 18° = \frac{GT}{AG}$$

$$\therefore \ GT = AG \times \tan 18°$$

$$= 11.2 \text{ m} \ (3 \text{ SF})$$

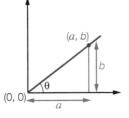

7. **(a)** Shaded area = area of sector − area of triangle

$$6.2 = \frac{1}{2}r^2\theta - \frac{1}{2}ab\sin C$$

$$6.2 = \frac{1}{2} \times 5^2\theta - \frac{1}{2} \times 5 \times 5\sin\theta$$

$$6.2 = \frac{25}{2}(\theta - \sin\theta)$$

$$\therefore \ \theta - \sin\theta = \frac{12.4}{25} = 0.496$$

(b) From GDC, $\theta = 1.49$ (3 SF)

8. **(a)** From the diagram:

$$\tan\theta = \frac{b}{a}$$

$$\Rightarrow b = a\tan\theta$$

(b) The gradient of the line is $\frac{b}{a} = \tan\theta$.

(c) The line makes angle θ with the x-axis, where $\tan\theta = $ gradient $= 3$. Also, $\theta \in [0, 90°]$.

Therefore, $\theta = \tan^{-1}(3) = 71.6°$.

9. **(a)** From the diagram:

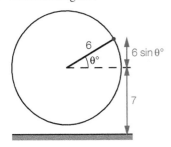

The height when $\theta = 50$ is $6 \times \sin 50° + 7 = 11.6$ m.

(b) The height at angle $\theta°$ is $h = 6\sin\theta° + 7$.

$$6\sin\theta° + 7 = 10$$

$$\Rightarrow \sin\theta° = \frac{1}{2}$$

$$\therefore \ \theta° = 30° \ \text{or} \ 180° - 30° = 150°$$

(c) From the graph below, $6\sin\theta° + 7 > 10$ when $30 < \theta < 150$.

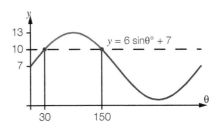

This corresponds to an angle of $120°$.

So the proportion of time the car is more than 10 m above the ground is $\frac{120°}{360°} = \frac{1}{3}$.

10. **(a)** $3x^2 - 2x + 5 = 3\left[x^2 - \frac{2}{3}x\right] + 5$

$$= 3\left[\left(x - \frac{1}{3}\right)^2 - \frac{1}{9}\right] + 5$$

$$= 3\left(x - \frac{1}{3}\right)^2 - \frac{1}{3} + 5$$

$$= 3\left(x - \frac{1}{3}\right)^2 + \frac{14}{3}$$

So the coordinates of the vertex are $\left(\frac{1}{3}, \frac{14}{3}\right)$.

(b) **(i)** $g(\theta) = 3\cos 2\theta - 4\cos\theta + 13$

$$= 3(2\cos^2\theta - 1) - 4\cos\theta + 13$$

$$= 6\cos^2\theta - 4\cos\theta + 10$$

(ii) Let $\cos\theta = x$. Then

$$6\cos^2\theta - 4\cos\theta + 10 = 6x^2 - 4x + 10$$

$$= 2(3x^2 - 2x + 5)$$

$$= 2f(x)$$

and $0 \leq \theta < 2\pi \Rightarrow -1 \leq x < 1$

Since the minimum value of $f(x)$ is $\frac{14}{3}$, the minimum value of $2f(x)$ is $\frac{28}{3}$.

That is, the minimum value of $g(\theta)$ is $\frac{28}{3}$.

11. (a) In triangle ABD:

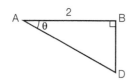

$$\cos\theta = \frac{2}{AD}$$

$$\Rightarrow AD = \frac{2}{\cos\theta}$$

Since AC is a diameter, $A\hat{E}C$ is a right angle.

So in triangle ACE:

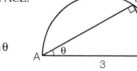

$$\sin\theta = \frac{CE}{3}$$

$$\Rightarrow CE = 3\sin\theta$$

Therefore

$$R = AD - CE$$

$$= \frac{2}{\cos\theta} - 3\sin\theta$$

$$= 2\sec\theta - 3\sin\theta$$

(b) For stationary values, $\dfrac{dR}{d\theta} = 0$.

$$\frac{dR}{d\theta} = 2\sec\theta\tan\theta - 3\cos\theta$$

$$0 = 2\sec\theta\tan\theta - 3\cos\theta$$

$$\Rightarrow 2\sec\theta\tan\theta = 3\cos\theta$$

$$\Rightarrow 2\times\frac{1}{\cos\theta}\times\frac{\sin\theta}{\cos\theta} = 3\cos\theta$$

$$\Rightarrow 2\sin\theta = 3\cos^3\theta$$

(c) From GDC, the minimum occurs at $\theta = 0.711$ and $R = 0.682$ (3 SF). So the smallest value of AD − CE is 0.682.

Going for the top 5

1.

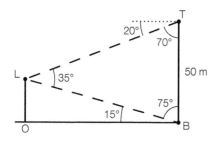

Since $O\hat{B}L = 15°$, $L\hat{B}T = 90° - 15° = 75°$.

Similarly, $L\hat{T}B = 90° - 20° = 70°$.

Therefore $B\hat{L}T = 180 - 70 - 75 = 35°$.

Using the sine rule in triangle BLT:

$$\frac{BL}{\sin70°} = \frac{50}{\sin35°}$$

$$\Rightarrow BL = \frac{50}{\sin35°}\times\sin70° = 81.915...$$

Then, in the right-angled triangle OBL:

$$\sin15° = \frac{OL}{81.915...}$$

$$\therefore\ OL = 21.2\text{ m}$$

2. Using the sine rule:

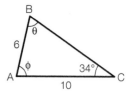

$$\frac{\sin A\hat{B}C}{10} = \frac{\sin34°}{6}$$

$$\Rightarrow \sin A\hat{B}C = \frac{10\sin34°}{6}$$

$$= 0.93198...$$

$$\sin^{-1}(0.93198...) = 68.746...$$

So $A\hat{B}C = 68.7°$ or $180 - 68.7° = 111.3°$.

If $A\hat{B}C = 68.7°$, the third angle $B\hat{A}C$ will be

$180° - 68.7° - 34°$ $= 77.3°$, which is fine.

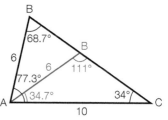

If $A\hat{B}C = 111.3°$, the third angle $B\hat{A}C$ will be

$180° - 111.3° - 34° = 34.7°$, which is also fine.

The area of the triangle is

$$\frac{1}{2}AB\times AC\times\sin B\hat{A}C = \frac{1}{2}\times6\times10\times\sin B\hat{A}C$$

$$= 30\times\sin B\hat{A}C$$

So the difference between the areas of the two possible triangles is

$$30\times\sin77.3° - 30\times\sin34.7° = 12.2\text{ cm}^2$$

3. (a) $\dfrac{1}{\tan x} - \tan x = 2$

$$\Leftrightarrow 1 - \tan^2 x = 2\tan x$$

$$\Leftrightarrow \tan^2 x + 2\tan x - 1 = 0$$

$$\Leftrightarrow \tan x = \frac{-2\pm\sqrt{2^2 - 4\times1\times(-1)}}{2\times1}$$

$$= \frac{-2\pm\sqrt{8}}{2}$$

$$= \frac{-2\pm2\sqrt{2}}{2}$$

$$= -1\pm\sqrt{2}$$

For $x\in\left[0, \dfrac{\pi}{2}\right[$, $\tan x \geq 0$, so only the positive value is possible. Hence $\tan x = -1+\sqrt{2}$.

(b) (i) $\dfrac{1}{\tan x} - \tan x = \dfrac{\cos x}{\sin x} - \dfrac{\sin x}{\cos x}$

$\qquad = \dfrac{\cos^2 x}{\sin x \cos x} - \dfrac{\sin^2 x}{\sin x \cos x}$

$\qquad = \dfrac{\cos^2 x - \sin^2 x}{\sin x \cos x}$

$\qquad = \dfrac{\cos 2x}{\frac{1}{2}\sin 2x}$

$\qquad = \dfrac{2}{\tan 2x}$

Alternatively,

$\dfrac{1}{\tan x} - \tan x$

$= \dfrac{1}{\tan x} - \dfrac{\tan^2 x}{\tan x}$

$= \dfrac{1 - \tan^2 x}{\tan x}$

$= 2\left(\dfrac{1 - \tan^2 x}{2\tan x}\right)$

$= 2\left(\dfrac{1}{\tan 2x}\right) \quad$ using $\tan 2x = \dfrac{2\tan x}{1 - \tan^2 x}$

$= \dfrac{2}{\tan 2x}$

(ii) $\dfrac{1}{\tan x} - \tan x = 2$

$\Leftrightarrow \dfrac{2}{\tan 2x} = 2$

$\Leftrightarrow \tan 2x = 1$

$x \in [0, \pi] \Rightarrow 2x \in [0, 2\pi]$

From the graph, there are two solutions to $\tan \theta = 1$ in $[0, 2]$:

One is $\tan^{-1}(1) = \dfrac{\pi}{4}$

And another is $\dfrac{\pi}{4} + \pi = \dfrac{5\pi}{4}$.

$2x = \dfrac{\pi}{4}, \dfrac{5\pi}{4}$

$\therefore x = \dfrac{\pi}{8}, \dfrac{5\pi}{8}$

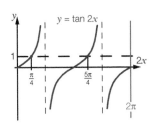

(c) From part (b)(ii), $x = \dfrac{\pi}{8}$ satisfies the equation

$\dfrac{1}{\tan x} - \tan x = 2$.

From part (a), if $\dfrac{1}{\tan x} - \tan x = 2$ and $x \in \left[0, \dfrac{\pi}{2}\right[$,

then $\tan x = -1 + \sqrt{2}$.

Therefore, $\tan \dfrac{\pi}{8} = -1 + \sqrt{2}$.

6 VECTORS

Mixed practice 6

1. If point D lies on the line, its position vector d satisfies

$$d = \begin{pmatrix} 1 \\ -1 \\ 5 \end{pmatrix} + \lambda \begin{pmatrix} -1 \\ 1 \\ -2 \end{pmatrix} = \begin{pmatrix} 1 - \lambda \\ -1 + \lambda \\ 5 - 2\lambda \end{pmatrix} \text{ for some value of } \lambda.$$

$$\therefore \overrightarrow{AD} = d - a = \begin{pmatrix} 1 - \lambda \\ -1 + \lambda \\ 5 - 2\lambda \end{pmatrix} - \begin{pmatrix} 3 \\ 2 \\ -1 \end{pmatrix} = \begin{pmatrix} -2 - \lambda \\ -3 + \lambda \\ 6 - 2\lambda \end{pmatrix}$$

For $\overrightarrow{AD}$ to be parallel to the x-axis,

$$\overrightarrow{AD} = \mu \begin{pmatrix} 1 \\ 0 \\ 0 \end{pmatrix} \quad \text{for some scalar } \mu$$

i.e. $\begin{pmatrix} -2 - \lambda \\ -3 + \lambda \\ 6 - 2\lambda \end{pmatrix} = \begin{pmatrix} \mu \\ 0 \\ 0 \end{pmatrix}$

$$\Rightarrow \begin{cases} -3 + \lambda = 0 \\ 6 - 2\lambda = 0 \end{cases}$$

Both equations are satisfied when $\lambda = 3$.

2. Vectors a and b are perpendicular if

$a \cdot b = 0$

$\sin\theta\cos\theta + \cos\theta\sin 2\theta = 0$

$\Leftrightarrow \sin\theta\cos\theta + \cos\theta(2\sin\theta\cos\theta) = 0$

$\Leftrightarrow \sin\theta\cos\theta(1 + 2\cos\theta) = 0$

$\Leftrightarrow \sin\theta = 0 \quad \text{or} \quad \cos\theta = 0 \quad \text{or} \quad \cos\theta = -\dfrac{1}{2}$

$\therefore \theta = 0, \pi, 2\pi, \dfrac{\pi}{2}, \dfrac{3\pi}{2}, \dfrac{2\pi}{3}, \dfrac{4\pi}{3}$

3. (a) $d = \overrightarrow{PQ} = q - p = \begin{pmatrix} -4 \\ 2 \\ 5 \end{pmatrix}$

Then, using $r = a + \lambda d$ with $a = p$ and the d found above, we get the vector equation

$$r = \begin{pmatrix} 3 \\ -1 \\ 2 \end{pmatrix} + \lambda \begin{pmatrix} -4 \\ 2 \\ 5 \end{pmatrix}$$

(b) $\overrightarrow{PM} = m - p = \begin{pmatrix} 0 \\ -3 \\ -1 \end{pmatrix}$

So the angle θ between l and (PM) satisfies

$$\cos\theta = \frac{d \cdot \overrightarrow{PM}}{|d||\overrightarrow{PM}|} = \frac{\begin{pmatrix} -4 \\ 2 \\ 5 \end{pmatrix} \cdot \begin{pmatrix} 0 \\ -3 \\ -1 \end{pmatrix}}{\sqrt{16+4+25}\sqrt{0+9+1}}$$

$$= \frac{-11}{\sqrt{45}\sqrt{10}}$$

$$\therefore \theta = \cos^{-1}\left(\frac{-11}{\sqrt{45}\sqrt{10}}\right) = 121.2°$$

But since we are asked for the acute angle, it is $180° - 121.2° = 58.8°$.

(c) Let the shortest distance from M to the line be x.

From the diagram,

$$\sin 58.8° = \frac{x}{|\overrightarrow{PM}|} = \frac{x}{\sqrt{10}}$$

$$\therefore x = \sqrt{10}\sin 58.8° = 2.70$$

4. (a) $\overrightarrow{BC} = c - b = \begin{pmatrix} -4-(-1) \\ 1-4 \\ 3-1 \end{pmatrix} = \begin{pmatrix} -3 \\ -3 \\ 2 \end{pmatrix}$

For ABCD to be a parallelogram, we must have $\overrightarrow{AD} = \overrightarrow{BC}$.

C(−4, 1, 3)   D

B(−1, 4, 1)   A(3, 0, 2)

$$\overrightarrow{AD} = d - a$$
$$\Rightarrow d = \overrightarrow{AD} + a$$
$$= \overrightarrow{BC} + a$$
$$= \begin{pmatrix} -3 \\ -3 \\ 2 \end{pmatrix} + \begin{pmatrix} 3 \\ 0 \\ 2 \end{pmatrix} = \begin{pmatrix} 0 \\ -3 \\ 4 \end{pmatrix}$$

i.e. coordinates of D are $(0, -3, 4)$.

(b) The equation of the diagonal [AC] is

$$r = \begin{pmatrix} 3 \\ 0 \\ 2 \end{pmatrix} + t\begin{pmatrix} -4-3 \\ 1-0 \\ 3-2 \end{pmatrix}$$

$$= \begin{pmatrix} 3 \\ 0 \\ 2 \end{pmatrix} + t\begin{pmatrix} -7 \\ 1 \\ 1 \end{pmatrix}$$

$$= \begin{pmatrix} 3-7t \\ t \\ 2+t \end{pmatrix}$$

The equation of diagonal [BD] is

$$r = \begin{pmatrix} -1 \\ 4 \\ 1 \end{pmatrix} + s\begin{pmatrix} 0-(-1) \\ -3-4 \\ 4-1 \end{pmatrix}$$

$$= \begin{pmatrix} -1 \\ 4 \\ 1 \end{pmatrix} + s\begin{pmatrix} 1 \\ -7 \\ 3 \end{pmatrix}$$

$$= \begin{pmatrix} -1+s \\ 4-7s \\ 1+3s \end{pmatrix}$$

These lines intersect when

$$\begin{cases} 3-7t = -1+s \\ t = 4-7s \end{cases}$$

$$\Leftrightarrow \begin{cases} 7t+s = 4 \\ t+7s = 4 \end{cases}$$

$$\Leftrightarrow s = \frac{1}{2} \text{ and } t = \frac{1}{2}$$

Substituting $t = \dfrac{1}{2}$ into the equation of [AC] gives the intersection point $(-0.5, 0.5, 2.5)$.

(Note that we know that the diagonals of a parallelogram must intersect, so there is no need to check the third equation.)

(c) Find the angle between the direction vectors

$$\overrightarrow{AC} = \begin{pmatrix} -7 \\ 1 \\ 1 \end{pmatrix} \text{ and } \overrightarrow{BD} = \begin{pmatrix} 1 \\ -7 \\ -3 \end{pmatrix}:$$

$$\cos\theta = \frac{\begin{pmatrix} -7 \\ 1 \\ 1 \end{pmatrix} \cdot \begin{pmatrix} 1 \\ -7 \\ -3 \end{pmatrix}}{\sqrt{49+1+1}\ \sqrt{1+49+9}}$$

$$= \frac{-7-7-3}{\sqrt{51}\sqrt{59}}$$

$$= -0.3099...$$

$$\therefore \theta = \cos^{-1}(-0.3099...) = 108.05°$$

So the acute angle is $180° - 108.05° = 71.9°$.

(d) $\overrightarrow{BA} \cdot \overrightarrow{BC} = \begin{pmatrix} 4 \\ -4 \\ 1 \end{pmatrix} \cdot \begin{pmatrix} -3 \\ -3 \\ 2 \end{pmatrix} = -12+12+2 = 2 \neq 0$

Adjacent sides are not perpendicular, so ABCD is not a rectangle.

5. (a) $\boldsymbol{p} \cdot \boldsymbol{q} = |\boldsymbol{p}||\boldsymbol{q}|\cos 60°$

$$\therefore x + 2x + 4 = \sqrt{1^2 + 2^2 + 2^2}\ \sqrt{x^2 + x^2 + 4} \times \frac{1}{2}$$

$$\Rightarrow 3x + 4 = \sqrt{9}\sqrt{2x^2 + 4} \times \frac{1}{2}$$

$$\Rightarrow 3x + 4 = \frac{3}{2}\sqrt{2x^2 + 4}$$

$$\Rightarrow (3x+4)^2 = \frac{9}{4}(2x^2 + 4)$$

$$\Rightarrow 9x^2 + 24x + 16 = \frac{9}{4}(2x^2 + 4)$$

$$\Rightarrow 18x^2 + 48x + 32 = 9x^2 + 18$$

$$\Rightarrow 9x^2 + 48x + 14 = 0$$

So $a = 9$, $b = 48$, $c = 14$.

(b) From GDC, $x = -0.3096...$ or $-5.0236...$

Let θ be the (acute) angle between vector $\boldsymbol{q}$ and the z-axis. Then

$$\cos\theta = \frac{\begin{pmatrix} x \\ x \\ 2 \end{pmatrix} \cdot \begin{pmatrix} 0 \\ 0 \\ 1 \end{pmatrix}}{\sqrt{x^2 + x^2 + 4}\ \sqrt{0^2 + 0^2 + 1^2}}$$

$$= \frac{2}{\sqrt{2x^2 + 4}}$$

$$= 0.977 \text{ or } 0.271$$

$$\therefore \theta = 12.3° \text{ or } 74.3°$$

6. (a) $\overrightarrow{BA} = \boldsymbol{a} - \boldsymbol{b} = \begin{pmatrix} 4-1 \\ 1-5 \\ 2-1 \end{pmatrix} = \begin{pmatrix} 3 \\ -4 \\ 1 \end{pmatrix}$

$$\overrightarrow{BC} = \boldsymbol{c} - \boldsymbol{b} = \begin{pmatrix} \lambda-1 \\ \lambda-5 \\ 3-1 \end{pmatrix} = \begin{pmatrix} \lambda-1 \\ \lambda-5 \\ 2 \end{pmatrix}$$

If there is a right angle at B, $\overrightarrow{BA} \cdot \overrightarrow{BC} = 0$.

i.e. $\begin{pmatrix} 3 \\ -4 \\ 1 \end{pmatrix} \cdot \begin{pmatrix} \lambda-1 \\ \lambda-5 \\ 2 \end{pmatrix} = 0$

$$\Leftrightarrow 3(\lambda-1) - 4(\lambda-5) + 1 \times 2 = 0$$

$$\Leftrightarrow 3\lambda - 3 - 4\lambda + 20 + 2 = 0$$

$$\Leftrightarrow \lambda = 19$$

(b) $\overrightarrow{AD} = 2\overrightarrow{DC}$

$$\Rightarrow \boldsymbol{d} - \boldsymbol{a} = 2(\boldsymbol{c} - \boldsymbol{d})$$

$$\Rightarrow 3\boldsymbol{d} = 2\boldsymbol{c} + \boldsymbol{a}$$

$$= 2\begin{pmatrix} 19 \\ 19 \\ 3 \end{pmatrix} + \begin{pmatrix} 4 \\ 1 \\ 2 \end{pmatrix} = \begin{pmatrix} 42 \\ 39 \\ 8 \end{pmatrix}$$

$$\Rightarrow \boldsymbol{d} = \begin{pmatrix} 14 \\ 13 \\ \frac{8}{3} \end{pmatrix}$$

So the coordinates of D are $(14, 13, \frac{8}{3})$.

7. (a) Equating the x, y and z components of l_1 and l_2:

$$\begin{cases} 3\lambda = 9 + 2\mu \\ 5\lambda = 15 \\ \lambda = 3 - \mu \end{cases}$$

From the second equation, $\lambda = 3$; then, from the third equation, $\mu = 0$.

Checking these values in the first equation: $3\lambda = 9 = 9 + 2\mu$, so the lines do intersect.

To find the coordinates of the point of intersection, substitute $\lambda = 3$ into the equation for l_1 (or substitute $\mu = 0$ into the equation for l_2) to get $(9, 15, 3)$.

(b) The angle between l_1 and l_2 is the angle between their direction vectors:

$$\cos\theta = \frac{\begin{pmatrix} 3 \\ 5 \\ 1 \end{pmatrix} \cdot \begin{pmatrix} 2 \\ 0 \\ -1 \end{pmatrix}}{\sqrt{9+25+1}\ \sqrt{4+0+1}}$$

$$= \frac{5}{\sqrt{35}\sqrt{5}}$$

$$\therefore \theta = \cos^{-1}\left(\frac{5}{\sqrt{35}\sqrt{5}}\right) = 67.8°$$

(c)   Velocity vector is $v = \begin{pmatrix} 3 \\ 5 \\ 1 \end{pmatrix}$ cm s^{-1}

Speed is the magnitude of velocity, so

$$\text{speed} = \left\|\begin{pmatrix} 3 \\ 5 \\ 1 \end{pmatrix}\right\| = \sqrt{9 + 25 + 1} = 5.92 \text{ cm s}^{-1}$$

(d)   At time t, the position of the

• first fly is $r = t\begin{pmatrix} 3 \\ 5 \\ 1 \end{pmatrix}$

• second fly is $r = \begin{pmatrix} 9 \\ 15 \\ 3 \end{pmatrix} + t\begin{pmatrix} 2 \\ 0 \\ -1 \end{pmatrix}$

From part (a), both flies pass through the point (9, 15, 3), but the first fly is there when $t = 3$ and the second one when $t = 0$. As there can be only one possible intersection point of these two straight line paths, the flies do not meet.

(e)   They are at the same height when their z-coordinates are the same:

$$t = 3 - t \Leftrightarrow t = 1.5$$

Substituting $t = 1.5$ into the two position vector expressions above gives:

• first fly: $r = t\begin{pmatrix} 3 \\ 5 \\ 1 \end{pmatrix} = \begin{pmatrix} 4.5 \\ 7.5 \\ 1.5 \end{pmatrix}$

• second fly: $r = \begin{pmatrix} 9 \\ 15 \\ 3 \end{pmatrix} + 1.5\begin{pmatrix} 2 \\ 0 \\ -1 \end{pmatrix} = \begin{pmatrix} 12 \\ 15 \\ 1.5 \end{pmatrix}$

i.e. the coordinates of the two flies are (4.5, 7.5, 1.5) and (12, 15, 1.5) when they are at the same height.

The distance between them is

$$d = \sqrt{(4.5 - 12)^2 + (7.5 - 15)^2 + (1.5 - 1.5)^2}$$
$$= \sqrt{112.5} = 10.6 \text{ cm}$$

(f)   (i)   The displacement vector from the first fly to the second fly is

$$\begin{pmatrix} 9 + 2t \\ 15 \\ 3 - t \end{pmatrix} - \begin{pmatrix} 3t \\ 5t \\ t \end{pmatrix} = \begin{pmatrix} 9 - t \\ 15 - 5t \\ 3 - 2t \end{pmatrix}$$

(ii)   The distance d between the flies at time t satisfies

$$d^2 = \left\|\begin{pmatrix} 9 - t \\ 15 - 5t \\ 3 - 2t \end{pmatrix}\right\|^2$$
$$= (9 - t)^2 + (15 - 5t)^2 + (3 - 2t)^2$$
$$= 315 - 180t + 30t^2$$

(iii)   Using a graph on the GDC (or by differentiation), the minimum value of d^2 is 45, so the minimum distance is $d = 6.71$ cm.

Going for the top 6

1.   (a)   As A lies on l_1, its position vector satisfies

$$a = \begin{pmatrix} 1 \\ 3 \\ 1 \end{pmatrix} + \lambda\begin{pmatrix} 1 \\ -1 \\ 2 \end{pmatrix} = \begin{pmatrix} 1 + \lambda \\ 3 - \lambda \\ 1 + 2\lambda \end{pmatrix} \text{ for some value of } \lambda.$$

As B lies on l_2, its position vector satisfies

$$b = \begin{pmatrix} 5 \\ -1 \\ -6 \end{pmatrix} + \mu\begin{pmatrix} 1 \\ 1 \\ 3 \end{pmatrix} = \begin{pmatrix} 5 + \mu \\ -1 + \mu \\ -6 + 3\mu \end{pmatrix} \text{ for some}$$

value of μ.

$$\therefore \overrightarrow{AB} = \begin{pmatrix} 5 + \mu \\ -1 + \mu \\ -6 + 3\mu \end{pmatrix} - \begin{pmatrix} 1 + \lambda \\ 3 - \lambda \\ 1 + 2\lambda \end{pmatrix} = \begin{pmatrix} \mu - \lambda + 4 \\ \mu + \lambda - 4 \\ 3\mu - 2\lambda - 7 \end{pmatrix}$$

Since $\overrightarrow{AB}$ is perpendicular to l_1,

$$\begin{pmatrix} \mu - \lambda + 4 \\ \mu + \lambda - 4 \\ 3\mu - 2\lambda - 7 \end{pmatrix} \cdot \begin{pmatrix} 1 \\ -1 \\ 2 \end{pmatrix} = 0$$
$$\Leftrightarrow (\mu - \lambda + 4) - (\mu + \lambda - 4) + 2(3\mu - 2\lambda - 7) = 0$$
$$\Leftrightarrow 6\mu - 6\lambda - 6 = 0$$
$$\Leftrightarrow \mu - \lambda = 1$$

(b)   Since $\overrightarrow{AB}$ is perpendicular to l_2,

$$\begin{pmatrix} \mu - \lambda + 4 \\ \mu + \lambda - 4 \\ 3\mu - 2\lambda - 7 \end{pmatrix} \cdot \begin{pmatrix} 1 \\ 1 \\ 3 \end{pmatrix} = 0$$
$$\Leftrightarrow (\mu - \lambda + 4) + (\mu + \lambda - 4) + 3(3\mu - 2\lambda - 7) = 0$$
$$\Leftrightarrow 11\mu - 6\lambda - 21 = 0$$
$$\Leftrightarrow 11\mu - 6\lambda = 21$$

(c) Solving the two equations in (a) and (b) simultaneously using the GDC gives $\mu = 3$, $\lambda = 2$

$$\therefore \overrightarrow{AB} = \begin{pmatrix} 3-2+4 \\ 3+2-4 \\ 9-4-7 \end{pmatrix} = \begin{pmatrix} 5 \\ 1 \\ -2 \end{pmatrix}$$

The shortest distance is $\left|\overrightarrow{AB}\right| = \sqrt{25+1+4} = \sqrt{30}$.

2. (a) $\overrightarrow{AB} = \dfrac{1}{2}\boldsymbol{a} + \dfrac{1}{2}\boldsymbol{b}$

    $\overrightarrow{BC} = \dfrac{1}{2}\boldsymbol{b} - \dfrac{1}{2}\boldsymbol{a}$

    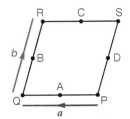

    (b) (i) $\boldsymbol{a}\cdot\boldsymbol{a} = |\boldsymbol{a}|^2$ and $\boldsymbol{b}\cdot\boldsymbol{b} = |\boldsymbol{b}|^2$. Since PQRS is a rhombus, $|\boldsymbol{a}| = |\boldsymbol{b}|$ and therefore $\boldsymbol{a}\cdot\boldsymbol{a} = \boldsymbol{b}\cdot\boldsymbol{b}$.

    (ii) $\overrightarrow{AB}\cdot\overrightarrow{BC} = \left(\dfrac{1}{2}\boldsymbol{a}+\dfrac{1}{2}\boldsymbol{b}\right)\cdot\left(\dfrac{1}{2}\boldsymbol{b}-\dfrac{1}{2}\boldsymbol{a}\right)$

    $= \dfrac{1}{4}\boldsymbol{a}\cdot\boldsymbol{b} - \dfrac{1}{4}\boldsymbol{a}\cdot\boldsymbol{a} + \dfrac{1}{4}\boldsymbol{b}\cdot\boldsymbol{b} - \dfrac{1}{4}\boldsymbol{b}\cdot\boldsymbol{a}$

    $= \dfrac{1}{4}(\boldsymbol{a}\cdot\boldsymbol{b} - \boldsymbol{b}\cdot\boldsymbol{a}) + \dfrac{1}{4}(\boldsymbol{b}\cdot\boldsymbol{b} - \boldsymbol{a}\cdot\boldsymbol{a})$

    $= \dfrac{1}{4}(\boldsymbol{a}\cdot\boldsymbol{b} - \boldsymbol{a}\cdot\boldsymbol{b})$ since $\boldsymbol{a}\cdot\boldsymbol{b} = \boldsymbol{b}\cdot\boldsymbol{a}$
    and $\boldsymbol{b}\cdot\boldsymbol{b} = \boldsymbol{a}\cdot\boldsymbol{a}$

    $= 0$

    Therefore $\overrightarrow{AB}$ and $\overrightarrow{BC}$ are perpendicular.

    (c) ABCD is a rectangle because $\overrightarrow{AB}$ and $\overrightarrow{BC}$ are perpendicular. (But since

    $\left|\overrightarrow{AB}\right|^2 = \left(\dfrac{1}{2}\boldsymbol{a}+\dfrac{1}{2}\boldsymbol{b}\right)\cdot\left(\dfrac{1}{2}\boldsymbol{a}+\dfrac{1}{2}\boldsymbol{b}\right)$ and

    $\left|\overrightarrow{BC}\right|^2 = \left(\dfrac{1}{2}\boldsymbol{b}-\dfrac{1}{2}\boldsymbol{a}\right)\cdot\left(\dfrac{1}{2}\boldsymbol{b}-\dfrac{1}{2}\boldsymbol{a}\right)$ are not

    necessarily equal, ABCD is not necessarily a square.)

3. (a) If $(-1, 5, 3)$ lies on l, then there exists a λ such that

    $$\begin{pmatrix} -1 \\ 5 \\ 3 \end{pmatrix} = \begin{pmatrix} -2 \\ 3 \\ 1 \end{pmatrix} + \lambda\begin{pmatrix} 1 \\ 2 \\ -1 \end{pmatrix}$$

    $$\begin{cases} -1 = -2+\lambda & \cdots (1) \\ 5 = 3+2\lambda & \cdots (2) \\ 3 = 1-\lambda & \cdots (3) \end{cases}$$

    $(1) \Rightarrow \lambda = 1$; $(2) \Rightarrow \lambda = 1$; $(3) \Rightarrow \lambda = -2$.

    There is no consistent value of λ, so A does not lie on l.

(b) Since B lies on l, its position vector is

$$\boldsymbol{b} = \begin{pmatrix} -2 \\ 3 \\ 1 \end{pmatrix} + \lambda\begin{pmatrix} 1 \\ 2 \\ -1 \end{pmatrix} = \begin{pmatrix} -2+\lambda \\ 3+2\lambda \\ 1-\lambda \end{pmatrix}$$ for some value of λ.

Therefore $\overrightarrow{AB} = \begin{pmatrix} -2+\lambda \\ 3+2\lambda \\ 1-\lambda \end{pmatrix} - \begin{pmatrix} -1 \\ 5 \\ 3 \end{pmatrix} = \begin{pmatrix} -1+\lambda \\ -2+2\lambda \\ -2-\lambda \end{pmatrix}$

Since $\overrightarrow{AB}$ is perpendicular to l,

$$\begin{pmatrix} -1+\lambda \\ -2+2\lambda \\ -2-\lambda \end{pmatrix}\cdot\begin{pmatrix} 1 \\ 2 \\ -1 \end{pmatrix} = 0$$

$\Leftrightarrow -1+\lambda - 4 + 4\lambda + 2 + \lambda = 0$

$\Leftrightarrow 6\lambda = 3$

$\Leftrightarrow \lambda = \dfrac{1}{2}$

Substituting back into the expression for $\boldsymbol{b}$ gives

$$\boldsymbol{b} = \begin{pmatrix} -2+\lambda \\ 3+2\lambda \\ 1-\lambda \end{pmatrix} = \begin{pmatrix} -1.5 \\ 4 \\ 0.5 \end{pmatrix}$$

So the coordinates of B are $(-1.5, 4, 0.5)$.

7 DIFFERENTIATION

Mixed practice 7

1. At a stationary point, $\dfrac{dy}{dx} = 0$

    $y = 2x - e^x$

    $\Rightarrow \dfrac{dy}{dx} = 2 - e^x$

    $0 = 2 - e^x$

    $\Leftrightarrow x = \ln 2$

    $x = \ln 2 \Rightarrow y = 2\ln 2 - e^{\ln 2} = 2\ln 2 - 2$

    So, the coordinates of the stationary point are $(\ln 2, 2\ln 2 - 2)$.

2. (a) Perimeter $= 2x + 2y$

    $40 = 2x + 2y$

    $\therefore y = 20 - x$

    Therefore,

    Area $= x(20-x)$

    $= 20x - x^2$

    (b) For maximum area, $\dfrac{dA}{dx} = 0$:

    $A = 20x - x^2 \Rightarrow \dfrac{dA}{dx} = 20 - 2x$

    $20 - 2x = 0 \Rightarrow x = 10$

$\dfrac{d^2A}{dx^2} = -2 < 0 \quad \therefore \text{ maximum}$

When $x = 10$, $y = 20 - x = 20 - 10 = 10$, i.e. a square.

3. $\dfrac{dy}{dx} = 5(\cos 3x)3 + 2x$

$\quad = 15\cos 3x + 2x$

When $x = \pi$, $y = 5\sin 3\pi + \pi^2 = \pi^2$ and

$\dfrac{dy}{dx} = 15\cos 3\pi + 2\pi = -15 + 2\pi$

Gradient of normal is $m = \dfrac{-1}{-15 + 2\pi} = \dfrac{1}{15 - 2\pi}$

So equation of the normal is:

$y - y_1 = m(x - x_1)$

$y - \pi^2 = \dfrac{1}{15 - 2\pi}(x - \pi)$

$\therefore y = \dfrac{1}{15 - 2\pi}x + \pi^2 - \dfrac{\pi}{15 - 2\pi}$

4.   (a) $s = at^2 + bt$

So $v = \dfrac{ds}{dt} = 2at + b \quad \cdots (*)$

and acceleration $= \dfrac{dv}{dt} = 2a$, a constant

(b)   Substituting the given information into (*):

$t = 1, v = 1 \Rightarrow 1 = 2a + b \quad \cdots (1)$

$t = 2, v = 5 \Rightarrow 5 = 4a + b \quad \cdots (2)$

$(2) - (1)$ gives $4 = 2a$

So $a = 2$

and then, from (1), $b = 1 - 2a = 1 - 4 = -3$.

5. $2x - y = 4$ can be rewritten as $y = 2x - 4$; the gradient is $m = 2$.

$y = 3\ln x + \dfrac{1}{x} = 3\ln x + x^{-1}$

$\Rightarrow \dfrac{dy}{dx} = \dfrac{3}{x} - x^{-2} = \dfrac{3}{x} - \dfrac{1}{x^2}$

For the tangent to be parallel to $2x - y = 4$, $\dfrac{dy}{dx} = 2$.

i.e. $\dfrac{3}{x} - \dfrac{1}{x^2} = 2$

$\Leftrightarrow 3x - 1 = 2x^2$

$\Leftrightarrow 2x^2 - 3x + 1 = 0$

$\Leftrightarrow (2x - 1)(x - 1) = 0$

$\Leftrightarrow x = \dfrac{1}{2}$ or $x = 1$

When $x = \dfrac{1}{2}$, $y = 3\ln\dfrac{1}{2} + 2$.

When $x = 1$, $y = 3\ln 1 + 1 = 1$.

Therefore, coordinates of the points are $\left(\dfrac{1}{2}, 3\ln\dfrac{1}{2} + 2\right)$ and $(1, 1)$.

6. $y = ax^3 + bx^2 + cx + d$

$\Rightarrow \dfrac{dy}{dx} = 3ax^2 + 2bx + c$

The graph has one stationary point if $3ax^2 + 2bx + c = 0$ has only one root, in which case the discriminant is zero:

$\Delta = (2b)^2 - 4(3a)c$

$0 = 4b^2 - 12ac$

$\therefore b^2 - 3ac = 0$

7.   (a) $h = 2t\sqrt{t} - \dfrac{1}{2}t^2 = 2t^{\frac{3}{2}} - \dfrac{1}{2}t^2$

So $v = \dfrac{dh}{dt} = 3t^{\frac{1}{2}} - t = 3\sqrt{t} - t$

(b)   The maximum value of v occurs when $\dfrac{dv}{dt} = 0$,

where $\dfrac{dv}{dt} = \dfrac{3}{2}t^{-\frac{1}{2}} - 1$.

$0 = \dfrac{3}{2}t^{-\frac{1}{2}} - 1$

$\Rightarrow t^{-\frac{1}{2}} = \dfrac{2}{3}$

$\Rightarrow t^{\frac{1}{2}} = \dfrac{3}{2}$

$\Rightarrow t = \dfrac{9}{4}$

At this time,

$v = 3\sqrt{\dfrac{9}{4}} - \dfrac{9}{4} = 3 \times \dfrac{3}{2} - \dfrac{9}{4} = \dfrac{9}{4} = 2.25 \text{ m s}^{-1}$

(c)   The rocket returns to the ground when $h = 0$ (for $t > 0$).

$2t\sqrt{t} - \dfrac{1}{2}t^2 = 0$

$\Rightarrow t\sqrt{t}\left(2 - \dfrac{1}{2}\sqrt{t}\right) = 0$

$\therefore 2 - \dfrac{1}{2}\sqrt{t} = 0 \quad$ since $t\sqrt{t} \neq 0$ (as $t > 0$)

$\Rightarrow \sqrt{t} = 4$

$\Rightarrow t = 16$

At this time, $v = 3\sqrt{16} - 16 = -4$, so the speed of the rocket is 4 m s^{-1}.

Notice that in part (b) we found that the maximum velocity was $2.25\,\mathrm{m\,s^{-1}}$. From part (c), the maximum *speed* is $4\,\mathrm{m\,s^{-1}}$, but this corresponds to a negative velocity (and $2.25 > -4$).

8. (a) The zeros are where $f(x) = 0$:

$x^4 - x = 0$

$\Leftrightarrow x(x^3 - 1) = 0$

$\Leftrightarrow x = 0$ or $x^3 - 1 = 0$

$\Leftrightarrow x = 0, 1$

(b) $f(x)$ is decreasing where $f'(x) < 0$:

$f'(x) < 0$

$\Leftrightarrow 4x^3 - 1 < 0$

$\Leftrightarrow x < \dfrac{1}{\sqrt[3]{4}}$

(c) $f''(x) = 12x^2$

$f''(x) = 0$

$\Leftrightarrow 12x^2 = 0$

$\Leftrightarrow x = 0$

(d) $f(x)$ is concave up where $f''(x) > 0$:

$f''(x) > 0$

$\Leftrightarrow 12x^2 > 0$

$\Leftrightarrow x^2 > 0$

$\Leftrightarrow x \in \left]-\infty, 0\right[\cup \left]0, \infty\right[$

(e) At a point of inflexion, $f''(x) = 0$, so from (c) the only possibility is $x = 0$. However, there must also be a change in concavity from one side of the point to the other; and since, by part (d), $f(x)$ is concave up on both sides of $x = 0$, this cannot be a point of inflexion.

(f)

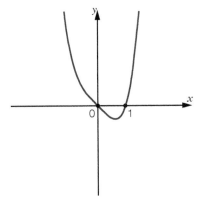

Going for the top 7

1. (a) $q = x^2$

(b) $d^2 = (x - 0)^2 + (q - 9)^2$

$= (x - 0)^2 + (x^2 - 9)^2$

$= x^2 + x^4 - 18x^2 + 81$

$= x^4 - 17x^2 + 81$

(c) The value of x that minimises d will also minimise d^2. At a minimum of d^2:

$\dfrac{\mathrm{d}}{\mathrm{d}x}(d^2) = 0$

$\dfrac{\mathrm{d}}{\mathrm{d}x}(x^4 - 17x^2 + 81) = 0$

$\Leftrightarrow 4x^3 - 34x = 0$

$\Leftrightarrow 2x(2x^2 - 17) = 0$

$\Leftrightarrow x = 0$ or $x = \pm\sqrt{\dfrac{17}{2}}$

To check which x value gives a minimum, consider the second derivative:

$\dfrac{\mathrm{d}^2}{\mathrm{d}x^2}(d^2) = 12x^2 - 34$

When $x = 0$, $\dfrac{\mathrm{d}^2}{\mathrm{d}x^2}(d^2) = -34 < 0$ $\quad \therefore$ maximum

When $x = \pm\sqrt{\dfrac{17}{2}}$,

$\dfrac{\mathrm{d}^2}{\mathrm{d}x^2}(d^2) = 12\left(\pm\sqrt{\dfrac{17}{2}}\right)^2 - 34 = 68 > 0$

$\therefore$ both minima

Hence, $x = \sqrt{\dfrac{17}{2}}, -\sqrt{\dfrac{17}{2}}$ give the minimum value of d^2.

(d) Therefore, the closest points to $(0, 9)$ on the curve $y = x^2$ are $\left(\sqrt{\dfrac{17}{2}}, \dfrac{17}{2}\right)$ and $\left(-\sqrt{\dfrac{17}{2}}, \dfrac{17}{2}\right)$.

2. (a) (i) If $p = 2$, the coordinates of P are $(2, e^2)$.

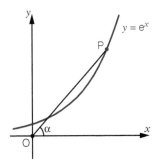

$\tan \alpha = \dfrac{e^2}{2} = 3.6945...$

$\therefore \alpha = \tan^{-1}(3.6945...) = 74.854... \approx 74.9°$

(ii) $y = e^x \Rightarrow \dfrac{dy}{dx} = e^x$

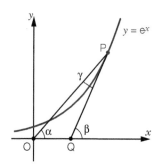

At P, the gradient of the tangent is e^2, so

$\tan \beta = e^2$

$\therefore \beta = 82.292... \approx 82.3°$

Using angles in the triangle OPQ:

$\alpha + (\pi - \beta) + \gamma = \pi$

$\Rightarrow \gamma = \beta - \alpha$

i.e. $\hat{OPQ} = 82.292... - 74.854... = 7.44°$.

(b) (i) At the point P: $y = e^p$ and $\dfrac{dy}{dx} = e^p$.
The equation of the tangent is:

$y - y_1 = m(x - x_1)$

$y - e^p = e^p(x - p)$

$\Rightarrow y = e^p x + e^p(1 - p)$

(ii) At Q, $y = 0$. So

$0 = e^p x + e^p(1 - p)$

$\Leftrightarrow e^p x = e^p(p - 1)$

$\Leftrightarrow x = p - 1$

Therefore, the coordinates of Q are $(p - 1, 0)$.

(c) If the tangent has equation $y = kx$, it must pass through the origin, which means that the coordinates of Q are $(0, 0)$. Hence $p = 1$.

The gradient of the tangent is k, so $k = e^p = e$.

3. (a) (i) $y = x^{-1} \Rightarrow \dfrac{dy}{dx} = -x^{-2}$

So at $x = p$, $\dfrac{dy}{dx} = -p^{-2} = -\dfrac{1}{p^2}$

i.e. the gradient of the tangent to the curve

at P is $-\dfrac{1}{p^2}$.

(ii) The y-coordinate of P is $y = \dfrac{1}{p}$.

Therefore, the equation of the tangent is:

$y - y_1 = m(x - x_1)$

$y - \dfrac{1}{p} = -\dfrac{1}{p^2}(x - p)$

$p^2 y - p = -(x - p)$

$p^2 y + x = 2p$

(b) (i) At Q, $x = 0$:

$p^2 y + 0 = 2p$

$\Rightarrow y = \dfrac{2}{p}$

So the coordinates

of Q are $\left(0, \dfrac{2}{p}\right)$.

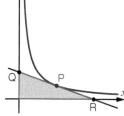

At R, $y = 0$:

$p^2 \times 0 + x = 2p$

$\Rightarrow x = 2p$

So the coordinates of R are $(2p, 0)$.

(ii) The area of triangle OQR is

$A = \dfrac{1}{2}bh = \dfrac{1}{2}(2p)\left(\dfrac{2}{p}\right) = 2$

(c) Since $Q = \left(0, \dfrac{2}{p}\right)$ and $R = (2p, 0)$,

$QR = \sqrt{(2p - 0)^2 + \left(0 - \dfrac{2}{p}\right)^2}$

$= \sqrt{4p^2 + \dfrac{4}{p^2}}$

$= 2\sqrt{p^2 + p^{-2}}$

(d) Sketching the graph of $d = 2\sqrt{p^2 + p^{-2}}$ using the GDC:

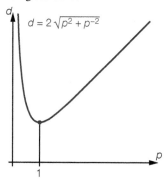

The minimum is at $p = 1$.

8 INTEGRATION

Mixed practice 8

1. (a) $\int \sqrt{e^x}\,dx = \int (e^x)^{\frac{1}{2}}\,dx$

$$= \int e^{\frac{x}{2}}\,dx$$

$$= 2e^{\frac{x}{2}} + c$$

(b) $\int_0^{\pi/10} 3\cos(5x)\,dx = \left[\frac{3}{5}\sin(5x)\right]_0^{\frac{\pi}{10}}$

$$= \frac{3}{5}\sin\left(\frac{\pi}{2}\right) - \frac{3}{5}\sin(0)$$

$$= \frac{3}{5}$$

2. Let $u = e^x + 1$; then $\dfrac{du}{dx} = e^x$, so $dx = \dfrac{1}{e^x}\,du$.

Changing the limits:

When $x = 0$, $u = e^0 + 1 = 2$.

When $x = \ln 2$, $u = e^{\ln 2} + 1 = 2 + 1 = 3$.

$$\int_0^{\ln 2} \frac{e^x}{\sqrt{e^x + 1}}\,dx = \int_{u=2}^{u=3} \frac{e^x}{\sqrt{u}}\frac{1}{e^x}\,du$$

$$= \int_{u=2}^{u=3} u^{-\frac{1}{2}}\,du$$

$$= \left[2u^{\frac{1}{2}}\right]_2^3$$

$$= 2\left(\sqrt{3} - \sqrt{2}\right)$$

3. First, sketch a graph of the curves to see where the intersection points are (and hence identify the region enclosed):

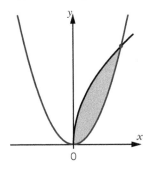

At the intersections:

$$\frac{1}{2}x^2 = 4\sqrt{x}$$

$$\Rightarrow x^2 = 8\sqrt{x}$$

$$\Rightarrow x^4 = 64x$$

$$\Rightarrow x(x^3 - 64) = 0$$

$$\Rightarrow x = 0 \text{ or } x^3 = 64$$

$$\therefore x = 0 \text{ or } 4$$

Therefore:

$$\text{Area} = \int_0^4 4x^{\frac{1}{2}} - \frac{1}{2}x^2\,dx$$

$$= \left[\frac{8}{3}x^{\frac{3}{2}} - \frac{1}{6}x^3\right]_0^4$$

$$= \left(\frac{8}{3}(4)^{\frac{3}{2}} - \frac{1}{6}(4)^3\right) - 0$$

$$= \frac{8}{3}(8) - \frac{64}{6}$$

$$= \frac{32}{3}$$

4. From GDC:

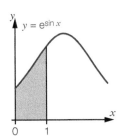

$$\int_0^1 e^{\sin x}\,dx = 1.63187$$

5. (a) $s = \int v\,dt = \int 3\sin t\,dt = -3\cos t + c$

When $t = 0$, $s = 0$, so $-3\cos 0 + c = 0 \Rightarrow c = 3$.

Therefore $s = -3\cos t + 3$.

When $t = \dfrac{3\pi}{2}$, $s = -3\cos\dfrac{3\pi}{2} + 3 = 3$ m.

(b) The maximum value of $s = -3\cos t + 3$ is achieved when $\cos t = -1$, i.e. when $t = \pi$ (and this maximum displacement is $s = 6$ m).

(c) The distance is equal to the area bounded between the velocity graph and the t-axis:

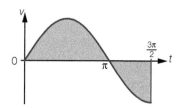

Distance travelled

$$= \int_0^\pi 3\sin t \, dt + \left| \int_\pi^{3\pi/2} 3\sin t \, dt \right|$$

$$= \left[-3\cos t \right]_0^\pi + \left| \left[-3\cos t \right]_\pi^{3\pi/2} \right|$$

$$= \left(-3(-1) - (-3)(1) \right) + \left| -3(0) - (-3)(-1) \right|$$

$$= (3+3) + |-3|$$

$$= 9 \text{ m}$$

6. $\int_0^a \sin 2x \, dx = \dfrac{3}{4}$

$$\Rightarrow \left[-\frac{1}{2}\cos 2x \right]_0^a = \frac{3}{4}$$

$$\Rightarrow \left(-\frac{1}{2}\cos 2a \right) - \left(-\frac{1}{2}\cos 0 \right) = \frac{3}{4}$$

$$\Rightarrow -\frac{1}{2}\cos 2a + \frac{1}{2} = \frac{3}{4}$$

$$\Rightarrow \cos 2a = -\frac{1}{2}$$

$$0 < a \le \pi \Rightarrow 0 < 2a \le 2\pi$$

$$\therefore 2a = \frac{2\pi}{3}, \frac{4\pi}{3}$$

$$\Rightarrow a = \frac{\pi}{3} \text{ or } \frac{2\pi}{3}$$

7. $y = \int \dfrac{2}{3x+1} \, dx = \dfrac{2}{3}\ln(3x+1) + c$

When $x = 0$, $y = 0$, so $0 = \dfrac{2}{3}\ln(0+1) + c \Rightarrow c = 0$.

$$\therefore y = \frac{2}{3}\ln(3x+1)$$

When $y = 2$:

$$2 = \frac{2}{3}\ln(3x+1)$$

$$\Rightarrow \ln(3x+1) = 3$$

$$\Rightarrow 3x+1 = e^3$$

$$\Rightarrow x = \frac{e^3 - 1}{3}$$

8. (a) $v = \int a \, dt$

$$= \int \frac{1}{(t-2)^2} \, dt = \int (t-2)^{-2} \, dt$$

$$= -(t-2)^{-1} + c = -\frac{1}{t-2} + c$$

When $t = 3$, $v = 5$, so $5 = -\dfrac{1}{3-2} + c \Rightarrow c = 6$.

$$\therefore v = -(t-2)^{-1} + 6$$

(b) $s = \int v \, dt$

$$= \int -(t-2)^{-1} + 6 \, dt$$

$$= -\ln(t-2) + 6t + C$$

When $t = 3$, $s = 20$, so $20 = -\ln(3-2) + 6(3) + C$.

$$\Rightarrow C = 2$$

$$\therefore s = -\ln(t-2) + 6t + 2$$

Therefore, when $t = 5$:

$$s = -\ln(5-2) + 6(5) + 2$$

$$= -\ln 3 + 32$$

$$= 30.9 \text{ m}$$

9. (a) $\cos 2\theta = 1 - 2\sin^2 \theta$

$$\Rightarrow 2\sin^2 \theta = 1 - \cos 2\theta$$

$$\Rightarrow \sin^2 \theta = \frac{1}{2}(1 - \cos 2\theta)$$

(b) $\int \sin^2 \theta \, d\theta = \int \dfrac{1}{2}(1 - \cos 2\theta) \, d\theta$

$$= \frac{1}{2}\left(\theta - \frac{1}{2}\sin 2\theta \right) + c$$

(c) Area of $R = \int_0^\pi \sin x \, dx$

$$= \left[-\cos x \right]_0^\pi$$

$$= (-\cos \pi) - (-\cos 0)$$

$$= -(-1) - (-1) = 1 + 1 = 2$$

(d) $V = \pi \int_0^\pi \sin^2 x \, dx$

$$= \pi \left[\frac{1}{2}\left(x - \frac{1}{2}\sin 2x \right) \right]_0^\pi \quad \text{(from part (b))}$$

$$= \frac{\pi}{2}\left[\left(\pi - \frac{1}{2}\sin 2\pi \right) - 0 \right]$$

$$= \frac{\pi^2}{2}$$

(e) Let $u = \cos x$; then $\dfrac{du}{dx} = -\sin x$ so $dx = -\dfrac{du}{\sin x}$.

$$\int \sin^3 x \; dx = \int \sin^2 x \sin x \; dx$$
$$= \int (1 - \cos^2 x)\sin x \; dx \quad \text{(using } \sin^2 x + \cos^2 x = 1)$$
$$= \int \sin x - \cos^2 x \sin x \; dx$$
$$= -\cos x - \int \cos^2 x \sin x \; dx$$
$$= -\cos x - \int u^2 \sin x \left(-\dfrac{du}{\sin x}\right)$$
$$= -\cos x + \int u^2 \; du$$
$$= -\cos x + \dfrac{1}{3}u^3 + c$$
$$= -\cos x + \dfrac{1}{3}\cos^3 x + c$$

Going for the top 8

1. (a) $x - 2 + \dfrac{C}{x+2} = \dfrac{(x-2)(x+2) + C}{x+2}$
$$= \dfrac{x^2 - 4 + C}{x+2}$$

Comparing this to $\dfrac{x^2 + 3}{x+2}$ gives $-4 + C = 3$, so $C = 7$.

(b) $\displaystyle\int \dfrac{x^2 + 3}{x+2} \; dx = \int x - 2 + \dfrac{7}{x+2} \; dx$
$$= \dfrac{1}{2}x^2 - 2x + 7\ln(x+2) + c$$

2. Let $u = \ln x$; then $\dfrac{du}{dx} = \dfrac{1}{x}$ so $dx = x \, du$.

Changing the limits:

When $x = e^{12}$, $u = \ln e^{12} = 12$.

When $x = e^3$, $u = \ln e^3 = 3$.

$\displaystyle\int_{e^3}^{e^{12}} \dfrac{1}{2x\ln x} \; dx = \int_3^{12} \dfrac{1}{2xu} x \, du$
$$= \int_3^{12} \dfrac{1}{2u} \; du$$
$$= \left[\dfrac{1}{2}\ln u\right]_3^{12}$$
$$= \dfrac{1}{2}(\ln 12 - \ln 3)$$
$$= \dfrac{1}{2}\ln 4$$
$$= \ln 4^{\frac{1}{2}}$$
$$= \ln 2$$

3. (a) From GDC:

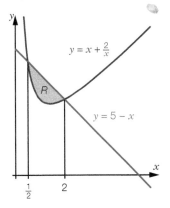

Points of intersection are $x = \dfrac{1}{2}$ and $x = 2$.

Area of $R = \displaystyle\int_{\frac{1}{2}}^{2} 5 - x - \left(x + \dfrac{2}{x}\right) dx$
$$= \int_{\frac{1}{2}}^{2} 5 - 2x - \dfrac{2}{x} \; dx$$
$$= \left[5x - x^2 - 2\ln x\right]_{\frac{1}{2}}^{2}$$
$$= (6 - 2\ln 2) - \left(\dfrac{9}{4} - 2\ln\dfrac{1}{2}\right)$$
$$= \dfrac{15}{4} - \ln 4 + \ln\dfrac{1}{4}$$
$$= 0.977$$

(b) $V = \pi\displaystyle\int_{\frac{1}{2}}^{2}(5 - x)^2 dx - \pi\int_{\frac{1}{2}}^{2}\left(x + \dfrac{2}{x}\right)^2 dx$
$$= \pi\int_{\frac{1}{2}}^{2} (5 - x)^2 - \left(x + \dfrac{2}{x}\right)^2 dx$$
$$= \pi\int_{\frac{1}{2}}^{2} 25 - 10x + x^2 - \left(x^2 + 4 + \dfrac{4}{x^2}\right) dx$$
$$= \pi\int_{\frac{1}{2}}^{2} 21 - 10x - \dfrac{4}{x^2} \; dx$$
$$= \pi\left[21x - 5x^2 + \dfrac{4}{x}\right]_{\frac{1}{2}}^{2}$$
$$= \pi\left[(42 - 20 + 2) - \left(\dfrac{21}{2} - \dfrac{5}{4} + 8\right)\right]$$
$$= \dfrac{27\pi}{4} = 21.2$$

4. Let $u = \sin\theta$

Then $\dfrac{du}{d\theta} = \cos\theta$, so $d\theta = \dfrac{du}{\cos\theta}$

$$\int \sin^4\theta\cos\theta \; d\theta = \int u^4 \cos\theta \dfrac{du}{\cos\theta}$$

$$= \int u^4 \, du$$

$$= \frac{1}{5}u^5 + c$$

$$= \frac{1}{5}\sin^5\theta + c$$

5. (a) $\displaystyle\int \cos^2 x - \sin^2 x \; dx = \int \cos 2x \; dx$

$$= \frac{1}{2}\sin 2x + c$$

(b) The point of intersection is the first positive x at which

$$\sin x = \cos x$$

$$\Leftrightarrow \frac{\sin x}{\cos x} = 1$$

$$\Leftrightarrow \tan x = 1$$

$$\therefore x = \frac{\pi}{4}$$

(c) Area of $R = \displaystyle\int_0^{\pi/4}\sin x \; dx + \int_{\pi/4}^{\pi/2}\cos x \; dx$

$$= \left[-\cos x\right]_0^{\pi/4} + \left[\sin x\right]_{\pi/4}^{\pi/2}$$

$$= \left(-\frac{1}{\sqrt{2}} - (-1)\right) + \left(1 - \frac{1}{\sqrt{2}}\right)$$

$$= 2 - \frac{2}{\sqrt{2}} = 2 - \sqrt{2}$$

(d) $V = \pi\displaystyle\int_0^{\pi/4}\cos^2 x \; dx - \pi\int_0^{\pi/4}\sin^2 x \; dx$

$$= \pi\int_0^{\pi/4}\left(\cos^2 x - \sin^2 x\right)dx$$

$$= \pi\left[\frac{1}{2}\sin 2x\right]_0^{\frac{\pi}{4}} \quad \text{(by part (a))}$$

$$= \pi\left[\frac{1}{2}\sin\frac{\pi}{2} - \frac{1}{2}\sin 0\right]$$

$$= \pi\left(\frac{1}{2} - 0\right) = \frac{\pi}{2}$$

9 DESCRIPTIVE STATISTICS

Mixed practice 9

1. (a) School 1: IQR $= 41 - 26 = 15$

School 2: IQR $= 48 - 34 = 14$

(b)

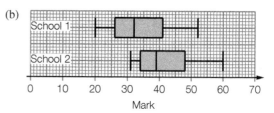

(c) The results have a similar spread. The results for School 2 are better on average.

2. (a) Using the formula $\bar{x} = \dfrac{\displaystyle\sum_{i=1}^{n} f_i x_i}{\displaystyle\sum_{i=1}^{n} f_i}$, the mean is

$$\frac{3\times 4 + 4\times 12 + 5x + 6\times 17 + 7\times 9}{4 + 12 + x + 17 + 9}. \text{ So}$$

$$\frac{12 + 48 + 5x + 102 + 63}{42 + x} = 5.23$$

$$\Leftrightarrow 225 + 5x = 5.23(42 + x)$$

$$\Leftrightarrow 225 - 219.66 = 0.23x$$

$$\Leftrightarrow x = \frac{5.34}{0.23}$$

$$\therefore x = 23 \quad \text{(as } x \text{ must be an integer)}$$

(b) There are 65 students, so the median is the 33rd number, which is 5.

3. (a) A (strong positive correlation)

(b) C (negative correlation)

(c) B (no linear correlation)

4. (a) From GDC, $r = 0.840$

(b) From GDC, $l = 3.61m + 38.0$

(c) For $m = 3.2$, predict $l = 3.61 \times 3.2 + 38.0 = 49.6\,\text{cm}$

(d) No; 5.6 is outside the range of the mass data (extrapolation is unreliable).

5. **(a)** Calculate the frequency table:

Age x (years)	Frequency
$16 \leq x \leq 26$	12
$26 < x \leq 36$	$46 - 12 = 34$
$36 < x \leq 46$	$82 - 46 = 36$
$46 < x \leq 56$	$90 - 82 = 8$

Then use this information to draw the histogram:

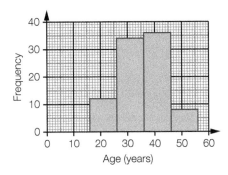

(b) Using the mid-interval values of the age brackets:

Mid-interval values	Frequency
21	12
31	34
41	36
51	8

From GDC, mean = 35.4 years and standard deviation = 8.31 years.

6. **(a)** Frequencies: 6, 8, 14, 10, 3, 5

(b) Using the mid-interval values of each class:

Distance (cm)	450.5	470.5	490.5	510.5	530.5	550.5
Frequency	6	8	14	10	3	5

From GDC, mean = 495 cm and standard deviation = 28.9 cm

(c) In inches:

$$\text{Mean} = \frac{495}{2.54} = 195 \text{ inches}$$

$$\text{Standard deviation} = \frac{28.9}{2.54} = 11.4 \text{ inches}$$

(d) Calculate the cumulative frequency up to each upper class boundary:

Distance (cm)	Frequency	Cumulative frequency
≤ 440	0	0
441–460	6	6
461–480	8	14
481–500	14	28
501–520	10	38
521–540	3	41
541–560	5	46

Then plot the cumulative frequency graph:

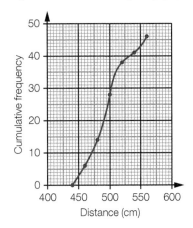

(e) **(i)** From the graph, a distance of 480 cm corresponds to a cumulative frequency of 14. So there are $46 - 14 = 32$ athletes who jumped more than 480 cm; this corresponds to a percentage of $\frac{32}{46} \times 100\% = 69.6\%$.

(ii) The probability of each athlete jumping further than 480 cm is $\frac{32}{46}$, so the probability that the two athletes both jumped more than 480 cm is $\frac{32}{46} \times \frac{32}{46} \approx 0.484$.

(f) 20% of 46 is 9.2, so the lower 80% of the athletes correspond to a cumulative frequency of $46 - 9.2 = 36.8$. From the graph in (c), a cumulative frequency of 36.8 corresponds to a distance of 515 cm. Therefore, the minimum distance required for qualification is 5.15 m.

1. Since there are 36 pieces of data,

$$9+13+x+y=36$$
$$\Rightarrow x+y=14 \quad \cdots (1)$$

Since the mean is $\dfrac{47}{9}$,

$$\frac{4\times 9+5\times 13+6x+7y}{9+13+x+y}=\frac{47}{9}$$
$$\Rightarrow 9(101+6x+7y)=47(22+x+y)$$
$$\Rightarrow 909+54x+63y=1034+47x+47y$$
$$\Rightarrow 7x+16y=125 \quad \cdots (2)$$

Solving (1) and (2) simultaneously gives $x=11$, $y=3$.

2. Among the known numbers there are two 3s and two 7s. Since the mode is 3, there must be another 3. Therefore at least one of x, y and z must be 3. Let us take $x=3$.

There are 11 numbers, so the median is the 6th number. As there are no 6s among the numbers known so far, one of the remaining unknowns must be 6. Let us take $y=6$.

The mean is 6, i.e.

$$\frac{3+2+3+7+10+5+7+12+3+6+z}{11}=6$$
$$\Rightarrow 58+z=66$$
$$\Rightarrow z=8$$

Therefore x, y and z are 3, 6 and 8.

3. (a) (i) $15+20+10=45$ students

(ii) 37 is two-fifths of the way from 35 to 40, so we can estimate that three-fifths of this group take more than 37 minutes; this corresponds to $\dfrac{3}{5}\times 10=6$ students.

(b) $\dfrac{6}{45}\approx 0.133$

4. (a) When $M=250$:

$$123+2.6a=250$$
$$\Rightarrow 2.6a=127$$
$$\Rightarrow a=48.85$$

So they need to spend $48.85.

(b) It is the number of members the club is expected to have if they spend no money on advertising.

5. (a) From the histogram, using mid-interval values:

Score	Frequency
45	10
55	20
65	30
75	30
85	20
95	10

From GDC: mean = 70, standard deviation = 13.8

(b) For standardised scores:

$$\text{Mean}=\frac{70-30}{10}=4$$

$$\text{Standard deviation}=\frac{13.8}{10}=1.38$$

(c) Within one standard deviation of the mean is between $70-13.8=56.2$ and $70+13.8=83.8$.

From the histogram, the number of students scoring between 56 and 84 is estimated to be

$$\left(\frac{4}{10}\times 20\right)+30+30+\left(\frac{4}{10}\times 20\right)=76$$

(d) A normal distribution has about 68% of data within one standard deviation of the mean.

$$\frac{76}{120}\times 100\%=63.3\%,$$ which is significantly less than 68%.

Therefore, the normal distribution is not a suitable model for these test scores.

10 PROBABILITY

Mixed practice 10

1. (a) Let $X=$ height of a tree. Then $X \sim N(26.2, 5.6^2)$.

$$P(X>30)=1-P(X\le 30)$$
$$=1-0.75129... \quad \text{(from GDC)}$$
$$=0.249$$

(b) Let $Y=$ the number of trees out of the 16 that are more than 30 m tall. Then $Y \sim B(16, 0.249)$.

$$P(Y\ge 2)=1-P(Y\le 1)$$
$$=1-0.06487... \quad \text{(from GDC)}$$
$$=0.935$$

2. Probability distribution of X:

x	1	2	3	4
$P(X=x)$	k	$4k$	$9k$	$16k$

$$k+4k+9k+16k=1 \Leftrightarrow k=\frac{1}{30}$$

$$E(X)=1(k)+2(4k)+3(9k)+4(16k)$$

$$=100k=\frac{100}{30}=3.33$$

3. Using a tree diagram with A first and letting $P(B \mid A)=x$:

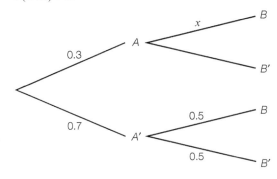

$$P(A \mid B)=\frac{P(A \cap B)}{P(B)}$$

$$=\frac{0.3x}{0.3x+0.7 \times 0.5}$$

But it is given that $P(A \mid B)=\frac{7}{16}$.

$$\therefore \frac{0.3x}{0.3x+0.35}=\frac{7}{16}$$

$$\Rightarrow 4.8x=2.1x+2.45$$

$$\Rightarrow x=\frac{49}{54}=0.907$$

4. $\sum p_i = 1$

$$\Rightarrow 0.3+p+q=1$$

$$\Rightarrow p+q=0.7 \quad \cdots (1)$$

$$E(Y)=3.1$$

$$\Rightarrow 0.1+0.4+3p+4q=3.1$$

$$\Rightarrow 3p+4q=2.6 \quad \cdots (2)$$

Solving equations (1) and (2) simultaneously gives $p=0.2$, $q=0.5$.

5. Listing the totals resulting from the possible outcomes on each die:

Total		Red die					
		1	1	1	4	5	6
Blue die	1	2	2	2	5	6	7
	2	3	3	3	6	7	8
	3	4	4	4	7	8	9
	4	5	5	5	8	9	10
	5	6	6	6	9	10	11
	6	7	7	7	10	11	12

Counting the relevant number of times a total occurs in each case:

(a) $P(\text{total greater than } 7)=\frac{12}{36}=\frac{1}{3}$

(b) $P(\text{total}=10 \mid \text{total} \neq 7)=\frac{3}{30}=\frac{1}{10}$

6. $P(\text{neither pink nor dress})$

$$=1-P(\text{pink} \cup \text{dress})$$

$$=1-\left(P(\text{pink})+P(\text{dress})-P(\text{pink} \cap \text{dress})\right)$$

Since the two properties are independent,

$$P(\text{pink} \cap \text{dress})=P(\text{pink}) \times P(\text{dress})$$

$$=0.3 \times 0.6 = 0.18$$

$$\therefore P(\text{neither pink nor dress})=1-(0.3+0.6-0.18)$$

$$=0.28$$

7. (a) Let X = number of requests in one day.

Some requests have to be turned down if there are more than 3 requests in a day.

$$P(X>3)=0.12+0.03=0.15$$

(b) $P(X=4 \mid X>3)=\dfrac{P\left((X=4) \cap (X>3)\right)}{P(X>3)}$

$$=\frac{P(X=4)}{P(X>3)}$$

$$=\frac{0.12}{0.15}=0.8$$

(c) Let Y = number of days in a seven-day week with more than 3 requests.

Then $Y \sim B(7, 0.15)$.

$$P(Y \geq 2)=1-P(Y \leq 1)$$

$$=1-0.71658... = 0.283$$

(d) Let N = number of vans hired out in one day.

When the number of requests is 3 or fewer, $N = X$; when the number of requests is more than 3, $N = 3$. Therefore:

$P(N = 0) = P(X = 0) = 0.07$

$P(N = 1) = P(X = 1) = 0.22$

$P(N = 2) = P(X = 2) = 0.35$

$P(N = 3) = P(X = 3) + P(X > 3) = 0.21 + 0.15 = 0.36$

So, the probability distribution of N is:

n	0	1	2	3
$P(N = n)$	0.07	0.22	0.35	0.36

(e) Let I = the income in one day; then $I = 120N$, so the probability distribution of I is:

i	0	120	240	360
$P(I = i)$	0.07	0.22	0.35	0.36

$E(I) = 0 \times 0.07 + 120 \times 0.22 + 240 \times 0.35 + 360 \times 0.36 = \$240$

(f) Let D = the distance travelled by a van.

Then $D \sim N(150, \sigma^2)$ and $P(D > 200) = 0.1$.

$Z = \dfrac{D - 150}{\sigma} \sim N(0, 1)$

$P\left(Z > \dfrac{200 - 150}{\sigma}\right) = 0.1$

$\Rightarrow \dfrac{200 - 150}{\sigma} = 1.28$ (from GDC)

$\Rightarrow \sigma = 39.0$ km

(g) $D \sim N(150, 39.0^2)$, so $P(D < 100) = 0.100$

P(two vans each travel less than 100 km)
$= 0.100^2 = 0.0100$

Going for the top 10

1.

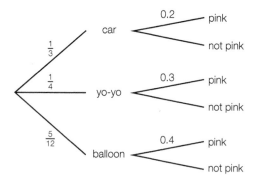

2. (a) Let $Z = \dfrac{X - \mu}{\sigma}$; then $Z \sim N(0, 1)$.

$P(X > 82) = 0.2$

$\Rightarrow P\left(Z > \dfrac{82 - \mu}{\sigma}\right) = 0.2$

$\Rightarrow \dfrac{82 - \mu}{\sigma} = 0.84162$

$\Rightarrow 82 - \mu = 0.84162\sigma$

$\Rightarrow \mu + 0.84162\sigma = 82$

(b) $P(X < 47) = 0.1$

$\Rightarrow P\left(Z < \dfrac{47 - \mu}{\sigma}\right) = 0.1$

$\Rightarrow \dfrac{47 - \mu}{\sigma} = -1.28155$

$\Rightarrow 47 - \mu = -1.28155\sigma$

$\Rightarrow \mu - 1.28155\sigma = 47$

Solving the two equations simultaneously (using GDC) gives $\mu = 68.1$, $\sigma = 16.5$.

3. (a) There are $12 + 20 + 18 = 50$ balls in total.

$P(\text{all 3 same colour})$
$= P(3 \text{ red OR } 3 \text{ green OR } 3 \text{ blue})$
$= P(3 \text{ red}) + P(3 \text{ green}) + P(3 \text{ blue})$
$= \dfrac{12}{50} \times \dfrac{11}{49} \times \dfrac{10}{48} + \dfrac{20}{50} \times \dfrac{19}{49} \times \dfrac{18}{48} + \dfrac{18}{50} \times \dfrac{17}{49} \times \dfrac{16}{48}$
$= 0.111$

(b) There are 6 different orders in which the 3 colours can be picked (RGB, RBG, GRB, GBR, BRG, BGR), and each of these has a probability of $\dfrac{12 \times 20 \times 18}{50 \times 49 \times 48} = 0.0367$.

Hence,

$P(\text{all 3 different colours}) = 6 \times 0.0367 = 0.220$

The worked solutions for the right column also include:

$P(\text{pink}) = \dfrac{1}{3} \times 0.2 + \dfrac{1}{4} \times 0.3 + \dfrac{5}{12} \times 0.4 = \dfrac{37}{120} = 0.308...$

$P(\text{balloon} \mid \text{pink}) = \dfrac{P(\text{balloon} \cap \text{pink})}{P(\text{pink})}$

$= \dfrac{\dfrac{5}{12} \times 0.4}{\dfrac{37}{120}} = \dfrac{20}{37} = 0.541$

(c) Using the conditional probability formula:

$$P(3\text{rd blue} \mid \text{first 2 blue}) = \frac{P(\text{all 3 blue})}{P(\text{first 2 blue})}$$

$$= \frac{\dfrac{18}{50} \times \dfrac{17}{49} \times \dfrac{16}{48}}{\dfrac{18}{50} \times \dfrac{17}{49}}$$

$$= \frac{16}{48} = \frac{1}{3} = 0.333$$

(Note that there is a more direct way to get to the answer: given that the first two balls are blue, there are then 16 blue balls left out of the total of 48. Hence, the probability that the third ball is also blue is $\dfrac{16}{48}$.)

(d) Listing all the ways that the third ball can be red, and adding their probabilities:

P(RRR or RGR or RBR or GRR or GGR
  or GBR or BRR or BGR or BBR)

$$= \frac{12 \times 11 \times 10}{50 \times 49 \times 48} + \frac{12 \times 20 \times 11}{50 \times 49 \times 48} + \frac{12 \times 18 \times 11}{50 \times 49 \times 48}$$

$$+ \frac{20 \times 12 \times 11}{50 \times 49 \times 48} + \frac{20 \times 19 \times 12}{50 \times 49 \times 48} + \frac{20 \times 18 \times 12}{50 \times 49 \times 48}$$

$$+ \frac{18 \times 12 \times 11}{50 \times 49 \times 48} + \frac{18 \times 20 \times 12}{50 \times 49 \times 48} + \frac{18 \times 17 \times 12}{50 \times 49 \times 48}$$

$$= \frac{6}{25} = 0.24$$

(more directly, we can argue that with no restrictions on the first two balls,

P(3rd is red) = P(1st is red)

$$= \frac{12}{50} = 0.24.)$$

4. (a) For Annie to win with her second shot, Annie, Brent and Carlos have to miss once and then Annie scores:

$$0.4 \times 0.5 \times 0.2 \times 0.6 = 0.024$$

(b) For Carlos to get a second shot, he has to miss once and Annie and Brent have to miss twice:

$$0.4^2 \times 0.5^2 \times 0.2 = 0.008$$

(c) (i) For Brent to win with his kth shot, he has to miss $(k-1)$ times and score once, Annie has to miss k times, and Carlos has to miss $(k-1)$ times:

$$0.5^{k-1} \times 0.5 \times 0.4^k \times 0.2^{k-1}$$

$$= (0.5 \times 0.4 \times 0.2)^{k-1} \times (0.5 \times 0.4)$$

$$= 0.04^{k-1} \times 0.2$$

(ii) P(Brent wins)

$$= \sum_{k=1}^{\infty} P(\text{Brent wins on } k\text{th shot})$$

$$= \sum_{k=1}^{\infty} 0.2 \times 0.04^{k-1}$$

Using the formula for the sum of the geometric series with $u_1 = 0.2$ and $r = 0.04$:

$$P(\text{Brent wins}) = \frac{0.2}{1 - 0.04} = 0.208$$

ANSWERS TO PRACTICE QUESTIONS

1 EXPONENTS AND LOGARITHMS

1. $\dfrac{\log 3}{\log 125}$

2. $\dfrac{\log 48}{\log\left(\frac{4}{9}\right)}$

3. $\dfrac{\log\left(\frac{8}{3}\right)}{\log 50}$

4. $x = 0$ or $x = 2$

5. $x = \ln 6$

6. $x = \ln 3$ and $y = \ln 2$, or $x = \ln 2$ and $y = \ln 3$

7. $\log\left(\dfrac{a^3 c}{b^2}\right)$

8. $1 + a + 2b - \dfrac{c}{2}$

9. $a = \dfrac{b^2}{10}$

10. $y = e^2 x^4$

11. (a) $y = \ln(800 - e^{2x})$ and $y = \dfrac{e^5}{x^3}$

    (b) $(2.89, 6.17)$, $(3.30, 4.12)$

12. $x = 8$

13. $x = 4 + 2\sqrt{6}$

14. $x = 2^{\pm\frac{1}{5}}$

15. $x = \dfrac{3}{2}$ or $-\dfrac{5}{4}$

16. (a) 2.13 (b) $2.37\,g$

17. $43\,105$

2 POLYNOMIALS

1. $x = 6 \pm \sqrt{2}$

2. $x = 2 \pm \sqrt{3}$

3. $x = 2k,\ 4k$

4. $x = \pm 1,\ \pm 2$

5. length $(\sqrt{59} - 3)$ m, width $(\sqrt{59} + 3)$ m

6. $k = \dfrac{1}{8}$

7. $k > 6$

8. $a = -5$ or 7

9. (a) $\left(x - \dfrac{9}{2}\right)^2 - \dfrac{65}{4}$ (b) $\left(\dfrac{9}{2}, -\dfrac{65}{4}\right) = (4.5, -16.25)$

10. (a) $3(x + 3)^2 - 7$ (b) $x = -3 \pm \sqrt{\dfrac{7}{3}}$

11. (a) $\left(x + \dfrac{b}{2}\right)^2 - \dfrac{b^2}{4} + c$

12. Since $(x - h)^2 \geq 0$, the smallest value of y will occur when $(x - h)^2 = 0$, i.e. when $x = h$. This value will be $y = 0 + k = k$. Hence the minimum point has coordinates (h, k).

13. $y = 4(x + 2)(x - 3) = 4x^2 - 4x - 24$

14. $y = -\dfrac{3}{4}(x - 4)(x + 2) = -\dfrac{3}{4}x^2 + \dfrac{3}{2}x + 6$

15. $b = -8,\ c = 22$

16. $y = -\dfrac{3}{4}(x - 4)(x - 8) = -\dfrac{3}{4}(x - 6)^2 + 3 = -\dfrac{3}{4}x^2 + 9x - 24$

17. (a) $b = -10,\ c = 24$ (b) $(4, 0)$ and $(6, 0)$

18. $61\,236$

19. (a) ± 2 (b) $560 x^4$

3 FUNCTIONS, GRAPHS AND EQUATIONS

1. $-3 < x < 3$

2. $x \in \mathbb{R},\ x \neq -1, 1$

3. $y \geq -8a^2$

4. $f^{-1}(x) = \dfrac{\ln x - b}{a}$

5. $x = \dfrac{5}{12}$

6. (a) 4 (b) 7

7.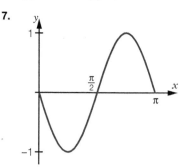

8. (a) Translation by vector $\begin{pmatrix} 2 \\ 0 \end{pmatrix}$ and vertical stretch with scale factor 3.

(b) Vertical stretch with scale factor 3 followed by translation by vector $\begin{pmatrix} 0 \\ -2 \end{pmatrix}$.

9.

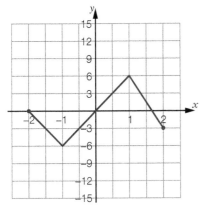

10.

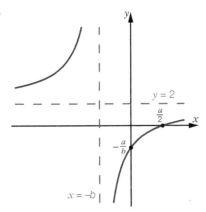

11. (a) $f^{-1}(x) = \dfrac{3x+1}{3-2x}$

(b) (i) Domain: $x \in \mathbb{R}, x \neq -1.5$. Range: $y \in \mathbb{R}$, $y \neq 1.5$

(ii) Domain: $x \in \mathbb{R}, x \neq 1.5$. Range: $y \in \mathbb{R}$, $y \neq -1.5$

12. (a)

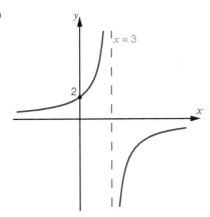

(b) $a = 3, b = 6$

13. $x = -1.23, 0.460$

14. $x = -2.46, -0.239, 1.70$

15. 1.53

16. $A = 5, k = 1$

17. $x = 5.61, y = 2.46$

18. $u_1 = 13, d = 3$

19. $a = 0.0208, b = 20.3$

20. (a) $26r + 42b = 272$ (b) 1.35 dollars

4 SEQUENCES AND SERIES

1. (a) -86 (b) 660

2. (a) 6.5 (b) 33rd term (c) 24 or 41

3. -0.669

4. 128 or 384

5. $a = 8, r = \dfrac{3}{2}$

6. (a) $a = \dfrac{1}{2}, r = 4$ (b) 9th term (c) 10

7. (a) $-3, 7$ (b) 54

8. (a) $\$32\,619$ (b) $\$1\,511\,552$

(c) 13th year (d) 27

9. (a) $\$185$ (c) 56

10. (a) $1.77147\,\mathrm{m}$ (b) 11th (c) $57\,\mathrm{m}$

11. (b) $k = 2.5$ (c) 28 years

5 TRIGONOMETRY

1. $a = 3, b = 2$

2. $f(x) \in \left[\dfrac{2}{7}, \dfrac{2}{3} \right]$

3. $\dfrac{3\pi}{4}$

4. $-\sqrt{\dfrac{5}{6}}$

5. $-\dfrac{\sqrt{7}}{4}$

6. $\dfrac{2}{\sqrt{5}}$

7. $x = \dfrac{\pi}{8}, \dfrac{7\pi}{8}$

8. $x = -\dfrac{8\pi}{9}, -\dfrac{5\pi}{9}, -\dfrac{2\pi}{9}, \dfrac{\pi}{9}, \dfrac{4\pi}{9}, \dfrac{7\pi}{9}$

9. $x = \dfrac{\pi}{36}, \dfrac{5\pi}{36}, \dfrac{25\pi}{36}, \dfrac{29\pi}{36}, -\dfrac{23\pi}{36}, -\dfrac{19\pi}{36}$

10. $x = \dfrac{\pi}{6}, \dfrac{\pi}{3}, \dfrac{2\pi}{3}, \dfrac{5\pi}{6}, \dfrac{7\pi}{6}, \dfrac{4\pi}{3}, \dfrac{5\pi}{3}, \dfrac{11\pi}{6}$

11. $\theta = -1.23, 1.23$

12. $x = \dfrac{2\pi}{3}, \dfrac{4\pi}{3}$

13. $x = 0, \pm\dfrac{\pi}{3}, \pm\pi$

14. $\theta = \dfrac{\pi}{6}, \dfrac{5\pi}{6}$

15. 6.75 cm

16. 23.2

17. 85.6°

18. 9.54 or 16.8

19. Area 154 cm²

6 VECTORS

1. (a) $\overrightarrow{MN} = \dfrac{1}{2}(\boldsymbol{c} - \boldsymbol{a})$ (b) $\overrightarrow{QP} = \dfrac{1}{2}(\boldsymbol{c} - \boldsymbol{a})$

2. (a) $\boldsymbol{r} = \begin{pmatrix} -3 \\ 1 \\ 2 \end{pmatrix} + \lambda \begin{pmatrix} 0 \\ 2 \\ 5 \end{pmatrix}$ (b) $(-3, 5, 12)$

3. The lines do not intersect (they are skew).

4. (a) $p = -9, q = 3$

  (b) $a = 4, b = -4$

5. (a) $\boldsymbol{r} = \begin{pmatrix} -1 \\ 3 \\ 3 \end{pmatrix} + \lambda \begin{pmatrix} 8 \\ -1 \\ -1 \end{pmatrix}$

  (b) $Q(23, 0, 0)$ (c) $\sqrt{594} \approx 24.4$

6. (a) No   (b) $\boldsymbol{r} = \begin{pmatrix} 1 \\ 4 \\ -3 \end{pmatrix} + t \begin{pmatrix} 2 \\ -1 \\ 1 \end{pmatrix}$

  (c) It does not.

7. 48.2°

8. 79.5°

9. (a) $\sqrt{33} \approx 5.74$ (b) 38.6°

10. (a) $\sqrt{53} \approx 7.28 \text{ ms}^{-1}$

  (b) 7.90°   (c) $2\sqrt{53} \approx 14.6 \text{ m}$

11. Speed $= 5\sqrt{17} \approx 20.6 \text{ km h}^{-1}$

7 DIFFERENTIATION

1. $\dfrac{dy}{dx} = 10x$

4. (a) $f'(x) = 2x - 2$

5. $8x\cos(3x) - 12x^2 \sin(3x)$

6. $\dfrac{9}{13}$

7. $2xe^{x^2} + \dfrac{3x\cos 3x - \sin 3x}{2x^2}$

8. $x = -4, 3$

9. $\dfrac{12 - 18x^2}{(x^2 + 2)^3}$

10. $y - e^{-12} = \dfrac{e^{12}}{12}(x - 2)$ or $e^{12}x - 12y + 12e^{-12} - 2e^{12} = 0$

11. -1.26

12. $(-1, 10)$ maximum, $(1, 6)$ minimum

13. $(0, 1)$ minimum, $\left(\dfrac{\pi}{2}, \dfrac{\pi}{2}\right)$ maximum,
  $\left(\dfrac{3\pi}{2}, -\dfrac{3\pi}{2}\right)$ minimum

14. $\ln(2\pi)$

15. Minimum area $= 48\pi$; $\dfrac{d^2A}{dr^2} = 2\pi + 256\pi r^{-3} > 0$
  when $r = 4$

16. 131 cm²

18. $(4, -121)$

19. $a = -12; b = -4$

20. (b) $a \geq -2$

  (d)

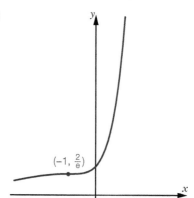

21.

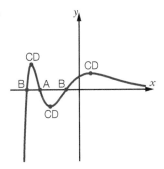

22. (a)

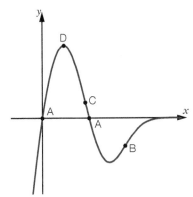

(b)   Concave down   (c)   Increasing

(d)   Minimum as $f'' > 0$

23. (a)   2 hours   (b) $(-)\dfrac{16\sqrt{3}}{9}$ km

(c) $8\,\text{km}\,\text{h}^{-1}$

24. (a) $v = \pi\cos\left(\dfrac{3\pi t}{2}\right), \; a = -\dfrac{3\pi^2}{2}\sin\left(\dfrac{3\pi t}{2}\right)$

(b) $t = \dfrac{2}{9}, \; \dfrac{4}{9}$

8 INTEGRATION

1. $2\sqrt{x} + \ln x + c$

2. $-\dfrac{1}{3}\cos 3x + \dfrac{1}{5}e^{5x} + \dfrac{2}{3}x^{\frac{3}{2}} + c$

3. $3x^3 + \dfrac{12}{5}x^{\frac{5}{2}} + \dfrac{1}{2}x^2 + c$

4. $\dfrac{1}{2}e^{2x} + 6e^x + 9x + c$

5. $x - \dfrac{1}{2}\cos 2x + c$

6. $-\dfrac{1}{2}\cos 2x + c$

7. $\dfrac{x^3}{3} - x - \dfrac{1}{4x} + c$

8. $\sqrt{2}$

9. 9

10. $\dfrac{2}{3}$

11. $\dfrac{\pi}{6}(e^{18} - e^6)$

12. $\dfrac{\pi}{2}$

13. (a) $(2, 4)$ (b) $\dfrac{416\pi}{15}$

14. (a) $5\left(t + \dfrac{1}{2}e^{-2t} - \dfrac{1}{2}\right)$

(b) $\dfrac{5}{4}(2t^2 - e^{-2t} - 2t + 1)$

(c) $\dfrac{5}{2}(9 + e^{-10})\,\text{m}\,\text{s}^{-1}$

15. $k = -25$

16. (a) $v = 10 - 2e^{-t}$ (b) $10.2\,\text{s}$

17. $\dfrac{e^{x^4}}{4} + c$

18. $e^{\sin x} + c$

19. $\dfrac{2}{3}(x^2 - 4)^{\frac{3}{2}} + c$

20. $-\ln(\cos x) + c$

21. $\ln(\sin x + 4) + c$

22. $\dfrac{1}{2}(\ln x)^2 + c$

9 DESCRIPTIVE STATISTICS

1. (a)   The values are already arranged in order and there are 13 of them, so the median is the 7th number. The lower quartile is therefore the median of the first six numbers, which is the mean of the 3rd and 4th numbers, i.e. the mean of 5 and 10.

(b)   14.5

2. $x = 2$

3. Median = 5, IQR = 2

4. (a) $k = 1$ (b)   1.38

5. (a) $x = 6, y = 7$ (b) $2.21\,\text{m}$

6. (a)   Lower: $23.5\,\text{m}$; upper: $29.5\,\text{m}$

(b)

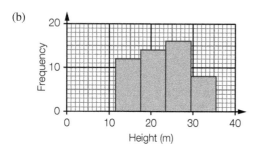

(c) Mean 22.9 m, standard deviation 6.12 m

(d) Mean 522.9 m, standard deviation 6.12 m

7. (a) Lower: 21; upper: 31; mid-interval value 26

(b)

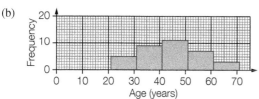

(c) Mean 44.3 years, standard deviation 11.6 years

8. (a)

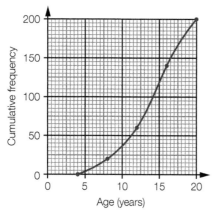

(b) 14   (c) 19

9. 12, 15, 10

10.

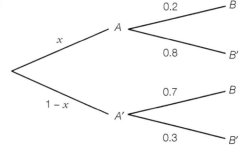

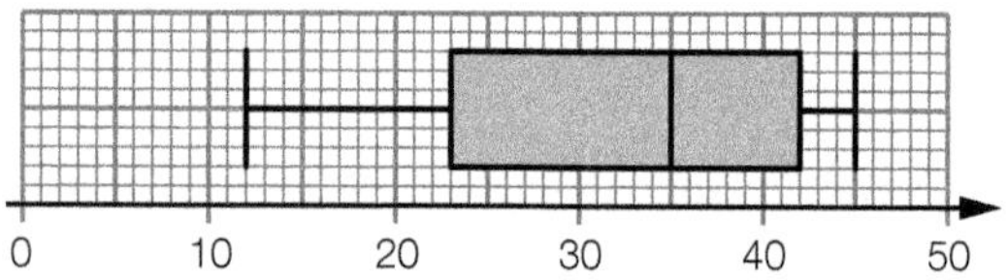

11. The average (median) times are the same. School B has larger spread of times. The distribution of times for school B is more symmetrical.

12. (a) $r = -0.355$. There is no correlation between the History and English marks.

(b) No, because there is no correlation between the two sets of marks.

13. (a) $r = -0.830$

(b) The independent variable is h; $t = -0.137h + 13.8$

(c) (i) 16 hours

(ii) It is reliable, because there is strong linear correlation and the value of t is within the range of the data.

(d) He is not necessarily right, because correlation does not imply causality.

14. (a) (i) 39 000

(ii) Not reliable because $A = 56$ is outside the range of the data.

(b) Not reliable because the correlation coefficient is close to 0 and so a linear model is not appropriate.

10 PROBABILITY

1. 0.33

2. (a) $\dfrac{9}{36} = 0.25$ (b) $\dfrac{14}{36} = 0.389$

3. (a) $\left(\dfrac{7}{8}\right)^{5}\left(\dfrac{1}{8}\right) = \dfrac{16\,807}{262\,144} \approx 0.0641$

(b) $\left(\dfrac{7}{8}\right)^{13} \approx 0.176$

4. 0.5

5. (a) 5   (b) $\dfrac{5}{7}$

6. (a)

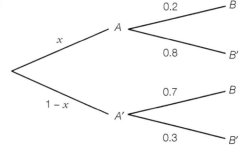

(b) 0.7   (c) No; $P(A) \neq P(A \mid B)$

7. $\dfrac{9}{23} \approx 0.391$

8. $\dfrac{1}{3}$

9. (a) $\dfrac{17}{30} \approx 0.567$ (b) 0.5

10. $E(Z) = 18.2$

11. $c = 0.2$, $p = 0.2$

12. 45.5

13. 0.0378

14. (a) The shots need to be independent; the probability of success on each shot needs to be constant.

    (b) 0.822

15. $p = \dfrac{2}{3}$, $n = 4$

16. (a) 0.233   (b) 0.202   (c) 7.04

17. 2.47 or 3.93

18. (a) 0.0668   (b) 0.653

19. 14.0 g

20. 152.8 ml

21. $\mu = 5.9$, $\sigma = 4.92$

11 EXAMINATION SUPPORT

Spot the common errors

1. (i) Tried to integrate a product factor by factor.

    (ii) Integrated e^{2x} to $2e^{2x}$ rather than $\dfrac{1}{2}e^{2x}$.

    (iii) Missed out '$+ c$'.

2. (i) Expanded $\ln(10 - x)$ into two logarithms.

    (ii) Sign error in performing the 'expansion' led to $-\ln x$, which then cancelled erroneously with the first $\ln x$.

    (iii) Tried to undo ln without first getting everything on each side of the equation (including minus signs) inside a ln.

    (iv) Obtained an answer which cannot go into the original expression (log of a negative number).

3. 'Cancelled' an expression which is not a factor of the entire denominator.